Titanium Alloys - Recent Progress in Design, Processing, Characterization, and Applications

Edited by Ram Krishna

Published in London, United Kingdom

Titanium Alloys - Recent Progress in Design, Processing, Characterization, and Applications
http://dx.doi.org/10.5772/intechopen.100680
Edited by Ram Krishna

Contributors
Yu Ren, Chaganti R.V.S. Nagesh, Ram Krishna, Yahya-Absalan, Behnaz Delaram, Faezeh Mokari, Mostafa Gholizadeh, Chonghe Li, Qisheng Feng, Alireza Farrokhi, Ilangovan Balasundar, Bipin Kedia

First published in London, United Kingdom, 2023 by IntechOpen
IntechOpen is the global imprint of INTECHOPEN LIMITED, registered in England and Wales, registration number: 11086078, 5 Princes Gate Court, London, SW7 2QJ, United Kingdom

British Library Cataloguing-in-Publication Data
A catalogue record for this book is available from the British Library

Additional hard and PDF copies can be obtained from orders@intechopen.com

Titanium Alloys - Recent Progress in Design, Processing, Characterization, and Applications
Edited by Ram Krishna
p. cm.
Print ISBN 978-1-80356-371-8
Online ISBN 978-1-80356-372-5
eBook (PDF) ISBN 978-1-80356-373-2

Meet the editor

Dr. Ram Krishna is an experienced metals researcher and academic with a background in metallurgy. He received his Bachelor of Engineering degree in metallurgical engineering and material science from the National Institute of Technology (NIT), Jamshedpur, and his Master's degree in materials and metallurgical engineering from the Indian Institute of Technology, Kanpur. Dr. Krishna continued his studies at the University of Leicester, UK, obtaining his Ph.D. in 2011. He has held postdoctoral positions at the University of Manchester, UK and North Carolina State University, USA. His research interests cover a wide range of materials for high-temperature applications, including mechanical testing of their behavior and performance, as well as advanced characterization techniques, as evidenced by his H-index of 13 and designation as co-inventor of 13 patents. He has authored or co-authored more than 35 research papers in journals and conference proceedings. His current research focuses on the 3D printing of metallic and ceramic alloys and materials for 3D printing applications.

Contents

Preface IX

Section 1
Design 1

Chapter 1 3
Titanium-Based Alloys with High-Performance: Design and Development
by Ram Krishna

Chapter 2 15
Titanium Alloys: Thermomechanical Process Design to Achieve Superplasticity in Bulk Material
by Bipin Kedia and Ilangovan Balasundar

Section 2
Extraction 35

Chapter 3 37
Titanium Extraction Metallurgy Developments and Control of Impurity Elements
by Chaganti R.V.S. Nagesh

Section 3
Processing 55

Chapter 4 57
Low-Cost Preparation Technologies for Titanium Alloys: A Review
by Qisheng Feng and Chonghe Li

Section 4
Mechanical Properties with Microstructural Considerations 81

Chapter 5 83
Shock-Induced Mechanical Response and Microstructure Evolution of Titanium Alloys
by Yu Ren

Chapter 6 **103**
Welding Properties of Titanium Alloys Grade 5
by Alireza Farrokhi

Section 5
Sustainable Use of Titanium Alloys **147**

Chapter 7 **149**
Use of the Advantages of Titanium in the Metal: Organic Framework
by Behnaz Delaram, Faezeh Mokari, Mostafa Gholizadeh and Yahya Absalan

Preface

Titanium and its alloys are known for their incredible versatility and use in a wide variety of applications. This is because of their high specific strength, biocompatibility, superior mechanical properties, corrosion resistance, and excellent properties at high temperatures, making them an excellent alternative to steel, copper alloys, nickel-based superalloys, and composite materials.

This peer-reviewed edited volume is specifically aimed at materials scientists, engineers, and technicians in the fields of technology development, production, design, and research in titanium and its alloys. It aims to enhance knowledge of advances in the development of alloys and their engineering properties. The book will serve as a useful reference for practicing engineers and researchers interested in lightweight materials, especially for aerospace, automobile, biomedical, pharmaceutical, power, oil, gas, and chemical industries applications.

The chapters cover a range of topics related to titanium extraction, processing, mechanical properties and applications. They are arranged into five sections: design, extraction, processing, mechanical properties with microstructural considerations and sustainable use of titanium alloys.

The first section (Design) discusses the major factors that need to be taken into consideration when designing titanium and its alloys. The second section (Extraction) deals with modern methods that can be used in the production of titanium and its alloys. The third section (Processing) deals with the problem of the low-cost processing of titanium and its alloys. In the fourth section (Mechanical Properties with Microstructural Considerations) we present the results of mechanical responses and microstructural evolution considerations with regard to titanium alloys, while the final section (Sustainable Use of Titanium Alloys) discusses titanium alloys in terms of their applications, including the possibility of sustainable use.

I would like to thank Prof. (Dr.) Karunesh Kumar Shukla, Director, NIT Jamshedpur and Dr. Ranjit Prasad, Head of the Metallurgical and Materials Engineering Department, NIT Jamshedpur for giving me the opportunity and time to carry out the editorial process of the book. I would appreciate any constructive suggestions from readers for further improvement.

Dr. Ram Krishna
Department of Metallurgical and Materials Engineering,
National Institute of Technology (NIT),
Jamshedpur, India

Section 1

Design

Chapter 1

Titanium-Based Alloys with High-Performance: Design and Development

Ram Krishna

Abstract

In recent years, titanium alloys with better properties have become increasingly popular. Their composition must be precisely designed to meet these demands. Screening alloy properties such as corrosion resistance, specific strength, properties to service at high temperatures, and microstructural stability requires a fair amount of effort and money to accomplish. By taking titanium-based alloys as an example, this chapter reviews the use of high-performance alloy design and development approach for industrial applications, in order to simplify the selection of titanium alloy compositions. The different high-throughput alloy design methods have been used by researchers to calculate diffusion coefficients of multiple elements using a thermodynamic database of atomic mobility. A composition with comprehensively optimal properties is selected by applying a rigorous screening criterion and then evaluating it in an experimental setting in order to come up with an optimal composition. Comparing this strategy with the data-driven material design methods that have been developed in recent times, few methods are more accurate and efficient, mainly because the diffusion pairs, the atomic mobility databases, and the refined physical models work together to make this strategy the most accurate and efficient. This approach could help develop high-performance titanium alloys, to overcome challenges of developing titanium alloys.

Keywords: titanium alloys, alloy design, high-throughput methods, microstructural stability, high-performance alloy

1. Introduction

Titanium alloys are used extensively for the manufacture of components used in automobile, chemical, aerospace, power generation, and biomedical applications that are subjected to complex operating conditions. It remains a challenge, however, to develop an alloy that has the desired combination of properties at an affordable cost. As an engineering material, they are useful in the manufacturing of turbine engines and aircraft components. The reason for this is that titanium alloys, among others, exhibit excellent strength at high-temperature applications, better creep resistance properties, good high-temperature microstructural stability, and resistance to corrosion and oxidation. The use of titanium alloys is not suitable for all parts of turbine engines due to phase equilibria and microstructural stability and these conventional titanium alloys cannot withstand operating temperatures greater than 600°C [1].

There is a range of advanced titanium alloys, which may provide titanium alloys with a higher temperature capability, such as TiAl, Ti3Al, and Ti2AlNb, as well as titanium/titanium aluminides [2]. This is a vital issue in terms of ensuring that these engineering materials maintain a high-temperature structural capability in order to be able to meet the increasing requirements for high thrust-to-weight ratios and energy efficiency that are currently being developed for turbine engine applications. These materials must be able to meet the increasing needs of high thrust-to-weight ratios and energy efficiency [3].

Titanium alloys have become more popular over the past few decades due to their ability to alloy with a wide variety of elements, such as Mo, Al, Ta, Zr, V, Zr, Mn, Fe, Ni, Co, Cr, Cu, and Nb [4]. There is no doubt that alloying elements play a crucial role in stabilizing either the low-temperature or high-temperature phases in titanium alloys [3, 5]. Depending upon the alloying elements in varying proportions in an alloy, they often result in the formation of low-temperature and high-temperature phases in the titanium alloy system. These phases included in the Ti-based alloy systems are defined as α, β, near-α, near-β, etc. Therefore, titanium alloys have been able to achieve excellent properties as a result of their chemical compositions [6]. There are three types of base titanium alloys mainly identified as α, β, and α + β alloys according to their phase stability.

Despite this, research has shown that there is a correlation between the overall properties of the alloy and the level of impurities it contains. Impurities adversely affect the plasticity of the alloy. There is a plastic deformation associated with titanium alloys when hydrogen, carbon, oxygen, and nitrogen combine with them [7]. As there are so many possible compositions and it is not possible to screen them in a practical manner using a random combination of these elements, there seems to be a compelling need for new approaches that will enable us to make efficient choices of compositions for the manufacturing of titanium alloys, and other advanced alloys of high performance. The interaction between the elemental composition of titanium alloys, their manufacturing techniques of them, and their microstructure of the alloys must be taken into account when determining the properties of titanium alloys and when they are being designed to get the best results. It is therefore imperative that one utilizes the correlation between the microstructure, properties, and performance of titanium alloys in order to gain a more comprehensive understanding of them. As a result of this correlation, it is expected that it will provide an opportunity to develop novel designs as a result of this correlation [8].

It has been demonstrated in this study that it will be feasible to use a new approach that will give correlations between the evolution of microstructure and properties of alloy systems and the interdiffusion properties of their compositional elements to arrive at a new approach to the problem. Evidence exists that indicates that titanium alloys obtain their strength as a result of the strengthening of the solid solution and grain refinement.

2. High-throughput materials characterization techniques

2.1 Mapping spatial data using statistical techniques

In order to characterize an alloy, the intrinsic heterogeneity of the material is utilized as a basis for a statistical spatial mapping technique that allows high throughput. There are tens of thousands of material microarrays that are being used to obtain different compositions, structures, and properties of a material through the cross-scale characterization of the material. It is necessary to formulate a statistical spatial-mapping model between the two sets of parameters based on the original material.

With the aid of high-throughput computational studies, it is possible to create databases by screening lattice units, which determine the properties of screened material with the help of high-throughput computations, and then creating a database of materials once the materials have been screened, and after the materials have been screened, the database can be created. There are numerous types of materials-design optimization strategies that have been developed to guide new materials discovery, process optimization, and material modification, as shown in **Figure 1**.

A Ti-alloy can have slightly different compositions, structures, as well as properties at different points of its structure, and the arrangement of these small differences determines the overall quality of the alloy as a whole, which is determined by the combination of these small differences. A wide range of rapid characterization techniques can be used to gather data from the macroscopic to the microscopic scale for the purpose of high-throughput statistical spatial mapping on a micron level. In order to meet practical sample sizes, fast and reasonable turnaround times are required. This is in order for composition, structure, and property datasets to be gathered at each of the locations. A database can be constructed based on the data entered into it, which contains spatial mapping lattices, based on precisely placed positional coordinates and references to point-to-point correspondence, in order to construct a spatial mapping map. The spatial mapping datasets are selected from a database within the required target intervals based on requirements for material research and development.

In order to determine a suitable design that is more likely to meet the targeted requirements based on a statistical analysis of the data, a statistical analysis can be performed in order to determine the appropriate design. A number of studies have

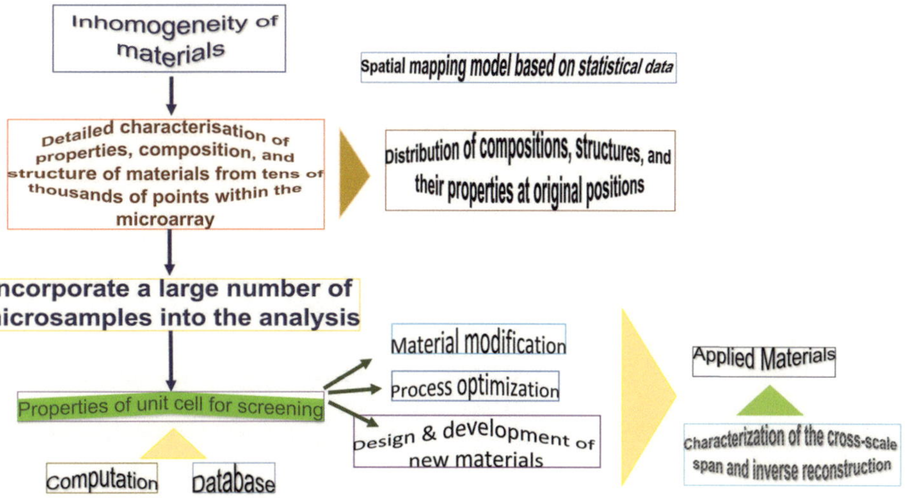

Figure 1.
It is a statistical spatial mapping technique based on the heterogeneity of the materials to generate the maps [9].

demonstrated that optimizing process parameters allows the assembly of these genetic units at the mesoscale to be verified, and quantitative correlations have been established between the micro-, meso-, and macroscales, as well as between practical samples and across the spectrum composition, structure, and properties. Recent years have seen the use of high-throughput statistical spatial mapping techniques to characterize a variety of material systems, including a wide variety of titanium alloys [10].

An alloy is a material that is heterogeneous, multielementary, and complex in structure. As a result, the structural composition, properties, and structure of an alloy may differ slightly at different points within it, and it is the amalgamation of these variances that determines the global functioning of the alloy. As a building block, a unit cell arrangement is used to provide insight into the properties of a material. The arrangement of unit cells is therefore critical for understanding the material and determining its properties at nanoscale. It is therefore possible to establish a correlation between the microscales, mesoscales, macroscales, and across-scale spans, as well as the compositional information, the structural information, and the properties of those spans, so as to enable the creation of novel materials and the amendment of current materials efficiently and economically. Rapid measurement of compositional information, structural information, and properties related to application at multilocations are performed in order to obtain practical sample sizes based on the available data. With the use of accurate positional coordinates, as well as point-to-point communication, a database is created to represent spatial mappings. Spatial mapping datasets are selected based on the target intervals as part of the design requirement of developing new materials. Based on a variety of factors, a variety of statistical analyses may be used to determine which design is best suited to meet the targeted requirements. Several criteria can be used to determine which design is best suited to meet the intended requirements, such as metrics and models that can be used to determine the frequency of occurrence within the range of parameters, the correlation ratio between parameters, and the statistical elimination of outliers. Many researchers used the process and used advanced microscopy and spectroscopy for data acquisition and statistical distribution analysis [11, 12].

2.2 Diffusion-multiple approach

In order to produce sizeable, multicomponents compositional deviations in alloy system samples through thermal interdiffusion, Zhao developed the diffusion multiple technologies, based on diffusion couples that generate compositional variations in bulk samples through diffusion [13]. Various experimental and analytical tools can be used to analyze diffusion multiples to extract the dependence of structure and properties on components. The application of the novel procedure significantly enhanced both the competence of elemental compositional design as well as the screening of appropriate heat treatment practices in comparison with the traditional methods that use a single alloy model to analyze the advancement rules of properties and microstructural information.

The infusion of a variety of alloying elements into titanium alloys can be investigated to determine how they affect their structure and properties by using combination of different diffusional multiple elements. It is, therefore, important and necessary to use diffusion multiple methods to study titanium alloys in order to achieve the best results [14].

It is possible to investigate kinetics, phase diagrams, and compositional-structural-properties relationships of alloy systems by using the diffusional multiple approaches, which uses the formation of compositional gradients and phase developed by long-term annealing of the alloy [15]. In order to determine diffusion coefficients and phase diagrams, conventional diffusion pairs and diffusion triples have been used for more than three decades. It has previously been demonstrated that it is possible to determine a number of composition-structure–property relationships by performing localized microscale property measurements on single-phase compositions [16].

Many systems have demonstrated the ability of the diffusion-multiple approach to be used as a tool for determining very complex phase diagrams, and this has been demonstrated for many different systems. In order to compare the phase diagrams of simple and very complex ternary systems, diffusion multiples have been used in place of equilibrated alloys to determine the phase diagrams [17]. In light of the results of this experiment, it can be concluded that the phase diagrams which have been determined from diffusion multiples are of very high accuracy [18].

As a result, a diffusion multiple analysis is a method that can be used to analyze diffusion data and to predict the microstructure and properties of alloy compositions using diffusion data for a variety of alloy compositions using the diffusion data as an input.

2.3 Computational thermodynamics using CALPHAD

CALPHAD is a computational thermodynamics program that can be used to compute and develop phase diagrams. It is commonly used for designing and developing new alloy systems [19]. In order to achieve the desired properties and consequently potential applications, it is important to examine the phase structure and phase equilibrium of the alloy systems. In the strategy of novel and advanced alloy systems, one of the advantages of using CALPHAD over entropy alone is its ability to analyze phase formation using free energy rather than entropy alone. This allows CALPHAD to better understand the function of system enthalpy, as well as the function of the entropy in the design of the alloy system [20].

A phase diagram provides detailed information on microstructural phase information as a function of its compositional information, its temperature information, and its information related to pressure. As such, it serves as a guide when designing and developing new Ti alloys. CALPHAD has been proven to be an effective tool for estimating the phases present in titanium alloy systems based on extensive research using titanium alloys. In spite of this, this method still did not yield enough screened titanium alloys which were able to produce the phases required in the temperature range of interest [21].

As a result of this, the thermodynamic databases in CALPHAD continue to grow as more and more experiments are conducted. Therefore, the accuracy of CALPHAD is also expected to increase as the number of experimental data continues to increase. It is expected that high-throughput CALPHAD simulations will be able to provide more accurate and reliable results for creating and optimizing titanium alloy compositions based on desired alloy properties because of more reliable and accurate simulation studies.

2.4 Machine learning and statistical methodology approach

This approach applies machine learning to a variety of different approaches, such as deep learning. This is a subset of machine learning, which is a subset of what we refer to as artificial intelligence. This is often referred to as artificial intelligence.

This refers to a computer system's capability of learning from the inputs it receives. This allows it to improve itself in order to find out how to do things better in the future. In the field of alloy design and development, artificial intelligence, including deep learning and machine learning algorithms, has emerged as a possible computational solution to overcome the challenges of alloy design and development, as well as control the costs and speed of processes in alloy design and development through artificial intelligence [14].

Recent years have seen a rise in interest in alloy development research centered on this approach. With the use of deep learning and machine learning algorithms, it will be feasible to rapidly transform large quantities of experimental data into usable feature information. With the aid of learning algorithms, it is possible to develop computer models that quickly generate judgment results based on input data. In addition, it is possible to addition, it is conceivable to extract information from alloy systems based on the prior knowledge of the system controllers.

Using high-throughput experiments and algorithms based on machine learning, Zhu et al. [14] have developed a titanium alloy using high-throughput experimental techniques. In the field of titanium-based alloy systems, the artificial neural network techniques, as traditional machine learning methods, have been successfully used for a variety of functions, such as predicting properties such as flow stresses, evolutions of microstructures, mechanical properties and parameters that affect during the processing of titanium alloys [22–24].

It has been discovered that when one technology is combined with machine learning, the screening of alloys becomes more efficient. A machine learning algorithm can be used to envisage the microstructure of an alloy and the anticipated results can be equated with those of the experimental results. Zhu et al. [14] reported the findings of a applied diffusion multiple in combination with machine learning algorithms to formulate a new Ti-based alloy system (Ti-3Al-2Nb-1.2 V-1Zr-1Sn-4Cr-4Mo). After the solution was heat-treated at 750°C for 6 hours and the material was aged at 550°C for 6 hours. In the study, researchers reported that better strength and plasticity could be obtained. The evidence suggests that the globular primary α phases elongated during deformation, while the secondary acicular α phases resist dislocation sliding, therefore, providing both a high degree of plasticity and strength for the alloys, that are subjected to the deformation [14].

3. Effect of alloying elements on properties

Various alloying elements are present in titanium alloys. The role of these alloying elements is to strengthen them either in low-temperature phase or in high-temperature phase, depending on the alloying elements. The alloying elements, in varying proportions, stabilize the close-packed hexagonal alpha (a) phase at low temperatures and the body-centered cubic beta (b) phase at high temperatures. It is their contents that determine the morphology and distribution of these phases. It is known that the alpha phase is a solid solution-strengthened phase that is stabilized by aluminum. This increases tensile and creeps strength. Tin is used in conjunction with aluminum to provide strength without embrittlement. Up to 5% of zirconium increases strength at low to intermediate temperatures. As the oxygen content in the titanium alloy increases, the ductility, toughness, and high-temperature strength of the titanium alloy decreases [25].

The high-temperature beta phase is stabilized by molybdenum, which increases the short-term strength at high temperatures. As well as being a beta-phase stabilizer, Niobium is also added to improve the stability of the surface at high temperatures. At all temperatures, silicon increases the strength as well as the creep resistance of titanium alloys. The other trace elements, such as chromium, cobalt, and nickel, are not beneficial for creep, and their contents are restricted to less than 0.01 percent [26].

Several studies have been conducted on the biocompatibility of titanium alloys for applications such as biomedical implants containing molybdenum, tantalum, and niobium, and on these bases, the developed alloys are Ti-Mo-Zr-Ta, Ti-12Mo-5Ta, and Ti-Nb-Zr-Mo [27, 28].

3.1 Case study on Ti-based alloy

It is important to realize that Ti-based alloy systems possess high-temperature mechanical properties, making them a very important group of structural materials that can be used in a wide range of strategic applications. These two-phase alloys are used for advanced engineering purposes, incorporating third alloying elements to enhance the ductility and strength, and maintain the properties at elevated temperatures. This is a two-phase lamellar structure consisting of alternate layers of tetragonal (L10) and hexagonal (D019) phases that consist of titanium and alloying elements in alternate layers. It is important to note that the optimum volume fraction for lamellar structure leads to an exceptional level of ductility that is virtually nonexistent in pure alloys. As a result of the process of plastic working, as well as the heat treatment, the microstructure of these alloys can be significantly altered in order to achieve a finely tuned mechanical property as well as fatigue behavior depending on the application [29]. There are a variety of mechanical properties depending on the morphology and the distribution of phases. As a result, the mechanical properties of titanium alloys with two phases are strongly influenced by the morphology of each phase. Many factors can affect the strength of an alloy with a lamellar microstructure; however, the thickness and diameter of the lamellae have the greatest impact [30]. In order to improve the mechanical properties of different alloys, the volume fractions, distribution, and morphology of the different phases play a critical role. Ti, Al, Cr, and Nb make up the elemental composition of the Ti-base alloy with nominal chemical compositions of Ti-40Al-2Cr-2Nb-0.4Y-0.2Zr, which has been used in this particular case. After one-hour heat treatment at 1350°C, the samples are furnace cooled to room temperature.

The engineering stress and strain and true stress–strain diagrams are shown in **Figure 2**. In fact, the true stress and strain values are very high because a smaller cross-sectional area is being used, whose section decreases continuously during elongation. True stress values indicate that, unlike engineering stress–strain values, material becomes stronger as strain is increased, compared to engineering stress–strain values. An alloy's mechanical properties can be greatly affected by the size of the colonies of crystallographically oriented lamellae within the alloy since it is a measure of the effective length of the slip that affects the alloy's mechanical properties. In spite of this, the transition to 'basket weave' microstructures will mean it will be even more challenging to determine the size of colonies as they emerge. Therefore, in order to illustrate the effect of microstructure refinement on mechanical properties, the thickness of lamellae was also taken into account as a quantitative parameter to illustrate the effect [31].

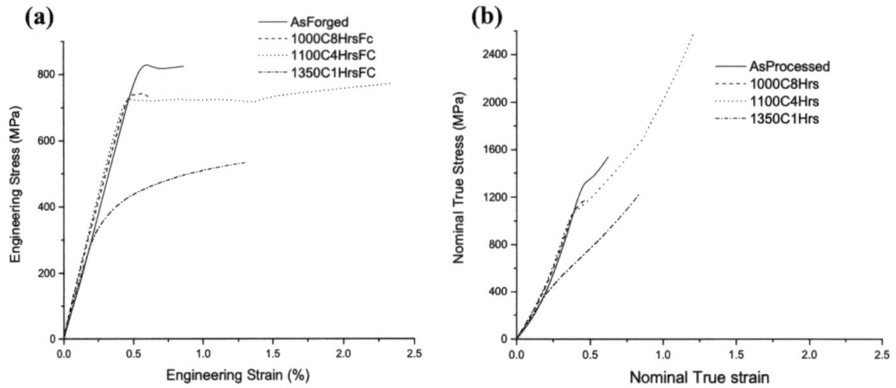

Figure 2.
A comparison of (a) engineering stress-strain curve and (b) true stress-strain curve of as-forged and solution heat-treated Ti-based alloys.

Strain hardening is the property of materials that exhibits this property. As part of the forming process, stain hardening (work hardening) plays an important role. Observing the plot, it was evident that stress rises without showing a drop in yield, as indicated in the figure. It can therefore be concluded from the shape of the true stress–strain curve that a material is prone to fracture before it is prone to yield, based on the shape of the curve.

The optical micrographs of as-forged and solution heat-treated Ti-alloy are shown in **Figure 3**. The lamellae structure of the annealed sample at 1350°C can be seen as having a random orientation due to the annealing process. This lamellae structure consists of alternate layers of the alloys γ- and α-phases. The solution heat-treated Ti-alloys has shown better property than forged alloys. This is due to the fact that the load-transferring capacity of lamellae is greater than that of duplex grains and near grains. A colony size of 80–100 μm was found to be the maximum size of the lamellae in the colony.

It can be seen in **Figure 4** that Ti-alloy has an even microstructure in an as-forged condition, which consists of equiaxed grains of γ and α phases and alternate plates of α and γ phases. Depending on the sample's history, the morphology of the grains differs from one sample to another. In the present case, the dislocations are thermal in origin. Several second-phase particles larger than 500 nm are usually found on the grain/interphase boundaries. It is worth noting that there is a wide variation in the

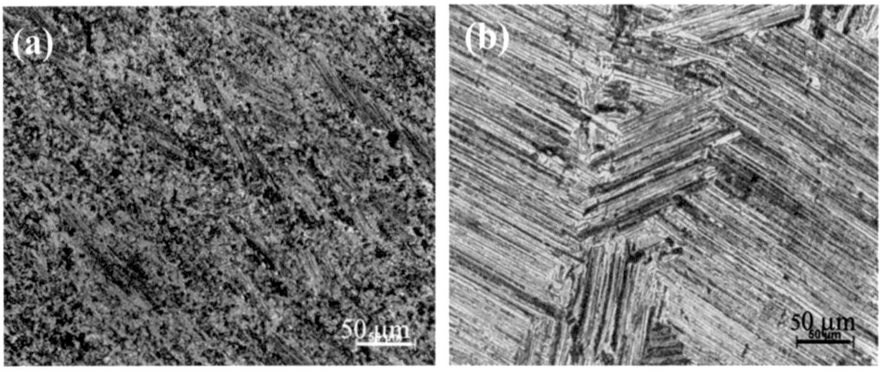

Figure 3.
Optical micrographs of (a) as-forged, and (b) solution-treated Ti-base alloy.

grain size in this multiphase microstructure. A few of the grains have a size of less than a micrometer, and there are a few others that are larger.

Figure 5 shows a dark field TEM micrograph of an alpha grain that is disordered. While the formation of α_2 from α is taking place, there are a number of finely ordered domains that are being formed, which are more apparent at the outset.

The deformation mechanism is also identified in the Ti-alloy. There is a high density of dislocations in the γ-phase, while there are very few dislocations in the α_2 phase. There is a great deal of difficulty in deforming the ordered alpha by dislocation slip. It is caused by the slitting of the dislocations, which causes them to become

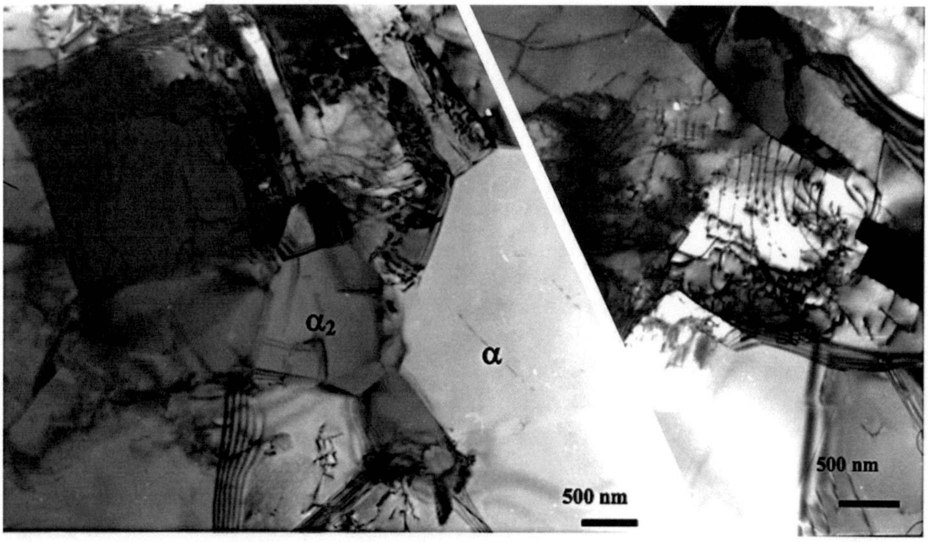

Figure 4.
Transmission electron microscope (TEM) micrographs of as-forged Ti-alloy in bright field mode showing equiaxed grains of γ and α phases and alternate plates of α and γ phases. The morphology of the phases depends on their history, or at least on the stage of their origin in the evolutionary process, which determines their morphology.

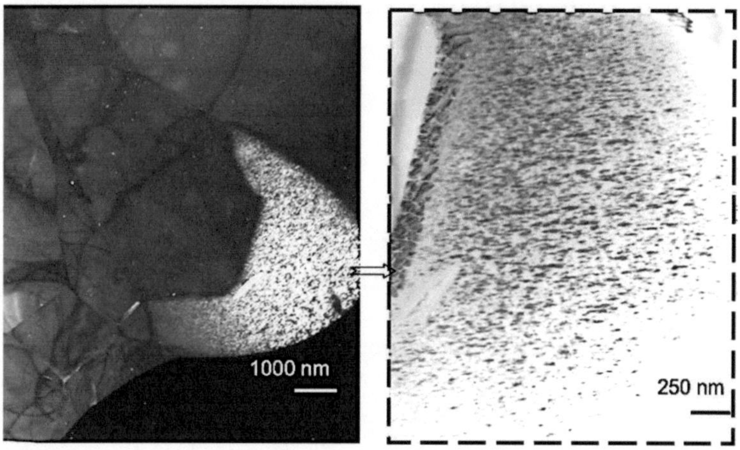

Figure 5.
A dark field TEM micrograph showing the transformation of $\alpha \rightarrow \alpha_2$ in disordered α grains showing the transformation in a blown out image. There are a number of the finely ordered domain during the formation of one α to another α_2, which are more evident at first.

super-dislocations. Super dislocations, as it is well known, require a greater amount of energy in order to move forward.

There is a deformation of the ordered α_2 phase by twinning. The disordered α phases are a high-temperature phase, which when cooled to room temperature decomposes to the ordered $\alpha_2 + \gamma$ at room temperature. A crucial aspect that we have collected in our research is the deformation characteristics of the constituent phases, which has provided us with invaluable information. A comparison should be made between the mechanical properties and microstructures of the selected alloy with a few other alloys that have been identified, and similar tests should be performed on those alloys that have been identified. By performing this comparative study, it will be easier to identify which sample is the best of those that have been tested. In addition, these alloys will also be able to provide an idea as to how to further improve the alloy design in Ti-base alloy systems by adding alloying elements or choosing the process of heat treatment, etc. which will result in better alloys.

4. Conclusions

It is evident from this article that the screening of alloy properties as well as microstructural stability is a substantial undertaking that requires a considerable amount of time and effort. In order to simplify the selection of titanium alloy compositions, a high-throughput-based alloy design approach is used. Different high-throughput methods have been used to calculate diffusion coefficients for a number of different elements, using a database of atomic mobility as a basis for calculating diffusion coefficients.

As a result of applying a rigorous screening criterion and evaluating it in an experimental setting in order to come up with the optimal composition, an optimal composition is selected that has comprehensively optimal properties. As compared to the data-driven materials design methods that have been used in recent years, few methods are more accurate and efficient, mainly because diffusion pairs, atomic mobility databases, and refined physical models work together so as to make this strategy the most accurate and efficient.

This approach is believed to be able to enable the development of high-performance titanium alloys regardless of the composition of the alloy, which is believed to be beneficial in overcoming the challenges that are associated with the development of novel titanium alloys for applications in high-temperature structural applications.

Author details

Ram Krishna
Department of Metallurgical and Materials Engineering, National Institute of Technology, Jamshedpur, India

*Address all correspondence to: krishnamme@gmail.com

References

[1] Leyens C, Peters M. Titanium and Titanium Alloys: Fundamentals and Applications. Weinheim, Wiley-VCH Verlag GmbH & Co. KGaA; 2003

[2] Santhosh R, Geetha M, Nageswara Rao M. Recent developments in heat treatment of Beta titanium alloys for aerospace applications. Transactions of the Indian Institute of Metals. 2017;**70**(7):1681-1688

[3] Meetham GW. The Development of Gas Turbines Materials. Dordrecht: Springer; 1981

[4] Peters M et al. Titanium alloys for aerospace applications. Advanced Engineering Materials. 2003;**5**(6):419-427

[5] Eylon D et al. High-temperature titanium alloys—A review. JOM. 1984;**36**(11):55-62

[6] Zhang T et al. In situ design of advanced titanium alloy with concentration modulations by additive manufacturing. Science. 2021;**374**(6566):478-482

[7] Barkia B et al. In situ monitoring of the deformation mechanisms in titanium with different oxygen contents. Materials Science and Engineering: A. 2015;**636**:91-102

[8] Zaefferer S. A study of active deformation systems in titanium alloys: Dependence on alloy composition and correlation with deformation texture. Materials Science and Engineering: A. 2003;**344**(1-2):20-30

[9] Haizhou Wang LZ, Jia Y, Li D, Yang L, Lu Y, Feng G, et al. State-of-the-art review of high-throughput statistical spatial-mapping characterization technology and its applications. Engineering. 2020;**6**:621-636

[10] Mao SS. High throughput growth and characterization of thin film materials. Journal of Crystal Growth. 2013;**379**:123-130

[11] Zhao L et al. A high-throughput strategy for rapid synthesis and characterization of Ni-based superalloys. Rare Metals. 2022;**41**:2693-2700

[12] Zhang X, Xiang Y. Combinatorial approaches for high-throughput characterization of mechanical properties. Journal of Materiomics. 2017;**3**(3):209-220

[13] Zhao J-CJ. The diffusion-multiple approach to designing alloys. Annual Review of Materials Research. 2005;**35**:51

[14] Zhu C et al. A titanium alloys design method based on high-throughput experiments and machine learning. Journal of Materials Research and Technology. 2021;**11**:2336-2353

[15] Zhao JC, Zheng X, Cahill DG. High-throughput diffusion multiples. Materials Today. 2005;**8**(10):28-37

[16] Zhao J-C. A combinatorial approach for efficient mapping of phase diagrams and properties. Journal of Materials Research. 2001;**16**(6):1565-1578

[17] Popp R et al. Determination of solubility limits of refractory elements in TCP phases of the Ni-Mo-Cr ternary system using diffusion multiples. Journal of Alloys and Compounds. 2019;**788**:67-74

[18] Zhao JC, Jackson MR, Peluso LA. Determination of the Nb–Cr–Si phase

diagram using diffusion multiples. Acta Materialia. 2003;**51**(20):6395-6405

[19] Gao MC, Alman DE. Searching for next single-phase high-entropy alloy compositions. Entropy. 2013;**15**(10):4504-4519

[20] Kumar J et al. Novel alloy design concepts enabling enhanced mechanical properties of high entropy alloys. Frontiers in Materials. 2022;**9**:868721

[21] Wu D et al. Designing high-strength titanium alloy using pseudo-spinodal mechanism through diffusion multiple experiment and CALPHAD calculation. Journal of Materials Science & Technology. 2021;**74**:78-88

[22] Luo J, Li M, Yu W. Prediction of flow stress in isothermal compression of Ti–6Al–4V alloy using fuzzy neural network. Materials & Design. 2010;**31**(6):3078-3083

[23] Sun Z, Yang H, Tang Z. Microstructural evolution model of TA15 titanium alloy based on BP neural network method and application in isothermal deformation. Computational Materials Science. 2010;**50**(2):308-318

[24] Sun Y et al. Determination of the influence of processing parameters on the mechanical properties of the Ti–6Al–4V alloy using an artificial neural network. Computational Materials Science. 2012;**60**:239-244

[25] Song Y et al. Theoretical study of the effects of alloying elements on the strength and modulus of β-type bio-titanium alloys. Materials Science and Engineering: A. 1999;**260**(1):269-274

[26] Bai W et al. Diffusivities and atomic mobilities in bcc Ti–Mo–Ta alloys. Calphad. 2022;**76**:102393

[27] Elias LM et al. Microstructural and mechanical characterization of biomedical Ti–Nb–Zr(–Ta) alloys. Materials Science and Engineering: A. 2006;**432**(1):108-112

[28] Marker C et al. Thermodynamic description of the Ti-Mo-Nb-Ta-Zr system and its implications for phase stability of Ti bio-implant materials. Calphad. 2018;**61**:72-84

[29] Sieniawski J et al. Microstructure and mechanical properties of high strength two-phase titanium alloys. In: Sieniawski J, Ziaja W, editors. Titanium Alloys: Advances in Properties Control. London: IntechOpen; 2013. pp. 69-80

[30] Gil FJ et al. The effect of cooling rate on the cyclic deformation of β-annealed Ti–6Al–4V. Materials Science and Engineering: A. 2003;**349**(1):150-155

[31] Ziaja W et al. Fatigue and microstructure of two-phase titanium alloys. Inzynieria Materialowa. 2001;**22**(5):981-984

Chapter 2

Titanium Alloys: Thermomechanical Process Design to Achieve Superplasticity in Bulk Material

Bipin Kedia and Ilangovan Balasundar

Abstract

Titanium alloys subjected to suitable thermomechanical processing (TMP) schedules can exhibit superplasticity. Most studies on superplasticity of titanium alloys are directed to sheet materials while studies on bulk materials are rather limited. Bulk Superplastic materials require lower load for forging aeroengine components. It further facilitates forming using non-conventional processes such as superplastic roll forming (SPRF). Multi axial forging (MAF), is employed here to achieve bulk superplasticity by imparting large strain without any concomitant change in external dimension. A comparison between uniaxial and MAF with respect to strain, strain path, initial microstructure and heat treatment was carried out to ascertain the microstructure refinement in Ti-6Al-4V alloy. A fine-grained structure was obtained after 3 cycles of MAF followed by static recrystallization at 850°C. Grain boundary sliding was observed in identified processing domain along with strain rate sensitivity (SRS) of 0.46 and maximum elongation of 815%. Validation of established ther¬momechanical sequence on a scaled-up work piece exhibited 640% elongation in domain (T = 820°C, $\dot{\varepsilon}$ = 3 x 10^{-4}/s) which indicated that the established TMP scheme can be used on a reliable and repeatable basis to achieve superplasticity in bulk material.

Keywords: titanium, Ti-6Al-4V, superplasticity, multi-axial forging, severe plastic deformation

1. Introduction

Titanium alloys are used extensively in the aerospace industry owing to their high specific strength, good static and dynamic properties, corrosion resistance, etc. [1, 2]. As an engine stator and rotor, they are used extensively in the compressor section as rings, discs and blades. A typical aeroengine compressor (low and high pressure) disc varies from 15 to 100 cm in diameter. These compressor discs are generally manufactured using either conventional or advanced forging techniques such as isothermal or near isothermal forging. Near α and α+β titanium alloys that are used in the compressor region of aeroengine exhibit a martensitic structure with coarse grains in the as-cast and homogenised condition [2]. By subjecting the material to thermomechanical processing (TMP), the lamellar structure can be converted into a fine-grained equiaxed structure. Numerous studies have been carried out to identify TMP parameters to break lamellar structure, and various theories have been put forward to describe the continuous dynamic recrystallisation (CDRX) or globularisation of α lamellae [1, 2]. It has been suggested that the formation of α/α interface (due to sub grain or shear band formation) in contact with β platelet gives rise to surface-driven penetration of the β phase into the lamellae leading to the breaking of lamellae into equiaxed structure. Thin lamellae structure has been reported to show better globularisation kinetics [3] due to easy penetration of β phase into α lamellae. It has been reported that high strain ($\varepsilon \geq 3$) is necessary to obtain a completely globularised structure [4] which could exhibit superplasticity.

Numerous studies have been carried out towards achieving superplasticity in various titanium alloy sheets [5–7]. However, work reported on realising superplasticity in the bulk material is very few. Multi-axial forging (MAF) was used to achieve submicron grain size in titanium alloy Ti-6Al-4V by Zherebstov et al. [8]. It was reported by Salishev et al. [4] that strain path change inherent in MAF process aids in increasing the globularisation kinetics or CDRX as multiple slip systems are activated during the strain path change. Poths et al. [9] subjected Ti-6Al-4V to monotonic and cyclic torsion in order to understand the effect of strain path on the globularisation kinetics of α lamellae. Based on the study and in contrast to the findings of Salishev et al. [4], Poths et al. [9] reported a decrease in kinetics of α globularisation and attributed the same to the change in strain path. Since contrasting results have been reported in literature and α lamellae globularisation is essential to achieve superplasticity in titanium alloys, it is imperative to carry out a systematic investigation on the influence of various factors that affect the globularisation kinetics of α lamellae. Further, as the objective is to achieve superplasticity in the bulk material, it is required to achieve this without modifying the external dimension of the material that facilitates subsequent secondary processing to produce the desired product/component.

A systematic study on the effect of strain, strain path, deformation temperature and starting or initial microstructure of Ti-6Al-4V on the globularisation or CRDX kinetics in a work horse titanium alloy Ti-6Al-4V is presented here. Further, a suitable thermomechanical process scheme that can maximise globularisation in the material on a reliable and repeatable basis is presented along with the temperature-strain rate regime under which the globularised material exhibits superplasticity.

2. Material

Ti-6Al-4V is an α+β alloy designed to provide moderately high strength, good fatigue strength and reasonable fracture toughness up to a temperature of 350°C. For aeroengine applications, the alloy is produced by vacuum arc remelting followed by thermomechanical processing in order to improve the structural integrity of the material. For the current study, triple vacuum arc remelted (VAR) titanium alloy Ti-6Al-4V ingot subjected to primary processing in the β and α+β field followed by mill annealing at 700°C for 1 h was procured from M/s M/s Mishra Dhatu Nigam, Hyderabad, India. The β transus of the 15 cm cylindrical mill annealed bars was reported to be 995 ± 5°C, and the same was reconfirmed through heat treatment experiments. Two billets of 15 cm diameter and 5 cm length were cut from the as-received mill annealed material. The billets extracted were coated with glass coating to prevent oxidation

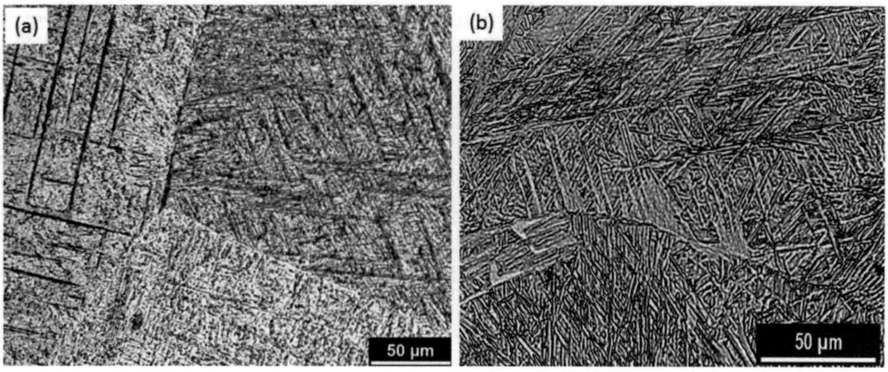

Figure 1.
Typical microstructure of Ti-6Al-4V subjected to heat treatment at 1015°C for 60 min followed by (a) water quenching – martensitic structure and (b) air cooling – lamellar structure.

during high-temperature exposure and were subjected to heat treatment above the β transus at a temperature of 1015°C. The billets were held at this temperature for a period of 60 minutes to achieve thermal homogeneity. One billet was then removed from the furnace and quenched in water while the second billet was cooled in air to obtain martensitic and lamellar microstructures respectively. Typical optical microstructures obtained after water quenching and air cooling are shown in **Figure 1a** and **b** respectively. The prior β grain size was estimated to be 601 + 86 μm and 684 + 64 μm for the water quenched and air-cooled material, respectively.

3. Effect of strain, strain path and microstructure

To evaluate effect of strain, strain path and initial microstructure on the globularisation kinetics of titanium alloy Ti-6Al-4V, both isothermal hot compression (uniaxial monotonic) and multi-axial (non-monotonic) compression experiments were carried out.

3.1 Isothermal hot compression

Cylindrical compression test samples with a constant height-to-diameter ratio of 1.5 were extracted from the water-quenched and air-cooled material using a wire-cut electro discharge machine (EDM) and lathe machine. The edges of the samples were chamfered to avoid fold formation during initial stages of deformation. The prepared samples were coated with glass (Deltaglaze 347) which acts as an oxidation resistor and lubricant. The coated samples were then heated to the desired test temperature and held at that temperature for 30 min in order to achieve thermal homogeneity. The cylindrical samples were then subjected to isothermal hot compression or uniaxial monotonic deformation (height reduction). A deformation of 25%, 40%, 58% and 78% was imparted to the samples at a constant true strain rate of 10^{-3}/s using a computer-controlled 200kN formability and workability testing machine custom built by M/s BISS, Bangalore. The deformation imparted corresponds to an equivalent strain of 0.29, 0.58, 0.87 and 1.51, respectively. After deformation, all the samples were quenched in water and were cut parallel to the deformation direction, hot-mounted and subjected to metallographic investigation following standard procedures. For stereological measurements of α fraction globularised, automatic and semi-automatic procedures were used and the α lamellae with an aspect ratio of <2.0 was considered to be globularised as reported in various literatures [4, 9]. Standard statistical measures such as relative accuracy (RA) and 95% confidence level [10, 11] were used to ensure reliability of stereology measurements. The microstructure of Ti-6Al-4V with an initial lamellar structure subjected to increasing amount of strain through isothermal hot compression, i.e. uniaxial monotonic deformation is shown in **Figure 2a–d**. Deformation of α lamellae and increased globularisation of α lamellae with increasing strain can be readily observed from **Figure 2**. Maximum globularisation was observed in the sample subjected to 78% deformation (**Figure 2d**) that corresponds to a strain of 1.51.

Conversion of lamellar structure of α phase into a globular morphology during deformation is considered to be a recrystallisation process, namely CDRX as against the discontinuous dynamic recrystallisation (DDRX) which has a distinctive nucleation and growth stage [12–14]. Globularisation of α lamellae present in the colony and at the grain boundary has been reported to take place by either sub grain or shear band formation. It was proposed by Margolin and Cohen [15] that subgrains form within the α lamellae during deformation followed by penetration of β phase into the α/α boundary with a simultaneous rotation of boundaries against each other resulting in coarsening of the recrystallised α when compared with the lamellae thickness from

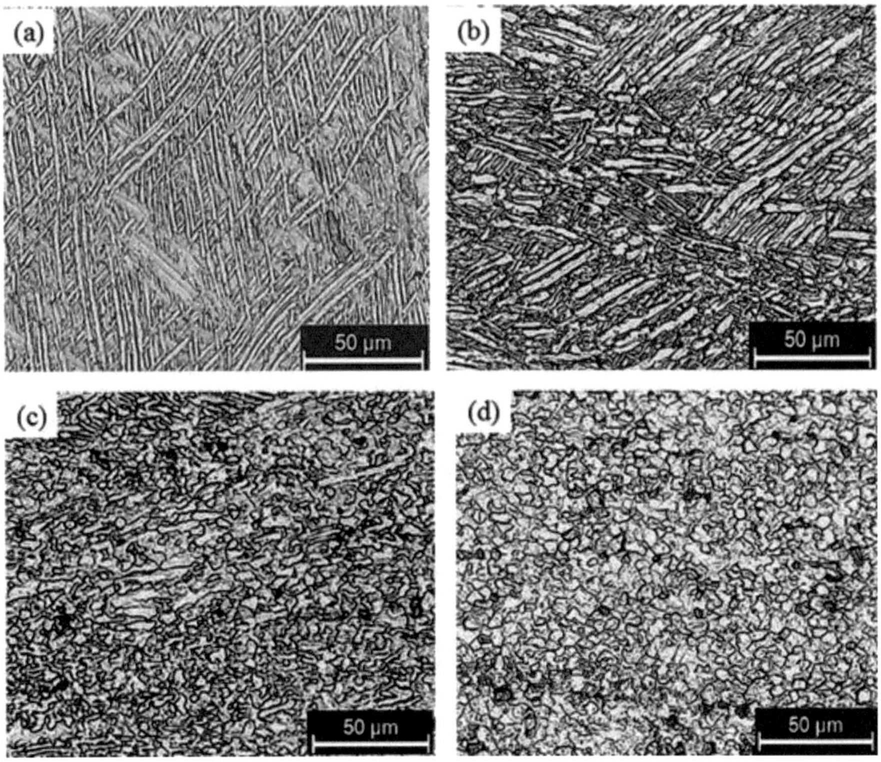

Figure 2.
Ti-6Al-4V with martensitic microstructure subjected to (a) 25 (b) 40 (c) 58 and (d) 78 percentage reduction or deformation at 900°C with a strain rate of 10^{-3}/s.

which it originated. Weiss et al. [3] reported formation of shear bands as main reason for CDRX of α lamellae. Irrespective of the mechanism, the formation of α/α interface in contact with β induces surface tension driven penetration of β phase resulting in globularisation. Balasundar [16] reported that both these mechanisms, namely sub grain and shear band formation operate in titanium alloys depending on the orientation of α lamellae and the processing conditions.

Based on the orientation of α lamellae with respect to the compression direction, bending and/or kinking of α lamellae can also be readily observed at different locations of the deformed sample. HCP α lamellae that have their c-axis aligned to the deformation direction require very high load or stress to deform [17] because such lamellae have a low Schmid factor and shear force along the slip direction. Though the globularisation fraction of α lamellae increases with increasing deformation, isolated regions where the α lamellae are still intact can be observed in the material even after imparting a stain of 151%. Though globularisation of grain boundary α lamellae could be observed after a stain of 58%, the prior β grain boundaries could be readily observed in the deformed material.

Microstructure of Ti-6Al-4V with an initial martensitic structure subjected to increasing amount of uniaxial monotonic deformation through isothermal hot compression is shown in **Figure 3a–d**. The influence of strain on the material with martensitic starting microstructure was observed to be similar to that in lamellar microstructure described above. However, the prior β grains were found to be destroyed completely by deformation, and it was not possible to identify them in the material. A quantitative discussion on the α fraction globularised is presented in Section 3.3.

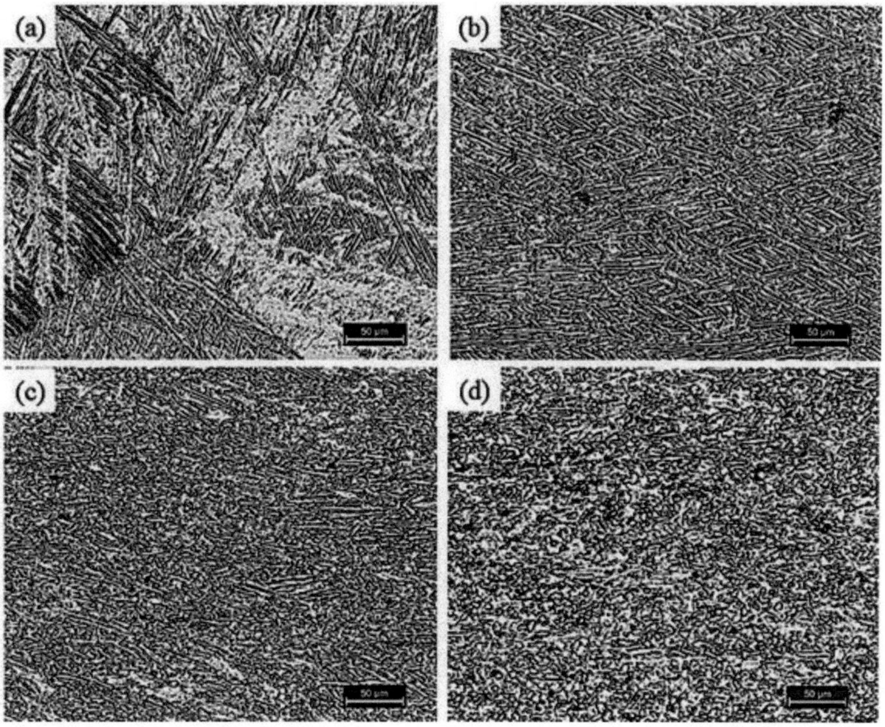

Figure 3.
Ti-6Al-4V with lamellar microstructure subjected to (a) 25 (b) 40 (c) 58 and (d) 78 percentage reduction or deformation at 900°C with a strain rate of 10^{-3}/s.

3.2 Multi-axial deformation

Conventional deformation process, such as forging, rolling, extrusion, etc., alters the dimension of the material that is being deformed. Severe plastic deformation is a metal working technique in which very high plastic strain can be imparted to the material without any concomitant change in the geometry or dimension. As the work piece material geometry is not altered, it provides an opportunity to deform the material repeatedly till the desired amount of strain is imparted. This large plastic strain results in the formation of ultrafine grain structure in the material. A large number of SPD processes are in vogue, MAF is one suchtechnique which is easier to implement as no special die or tooling is required for deforming the material. In MAF, the material is deformed in cyclic way along all the three orthogonal directions. Each MAF cycle consist of three deformation steps, i.e. imparting equal amount of deformation along the three orthogonal directions (X, Y, Z) as shown in **Figure 4**.

The strain path is altered when the sample is rotated by 90° during MAF. This change in strain path is expected to assist in refining the grain size. After three processing or deformation steps, i.e. after a cycle of MAF, the work piece reverts back to its original dimension. Since the dimension of the work piece remains the same, it is possible to carry out MAF cycles multiple times and thereby impart a large amount of strain to the material as desired or till the failure of material. The amount of strain imparted for a given percentage of reduction and the number of MAF cycles can be calculated as per the following relation:

$$\varepsilon_{eq} = 3N\ln(1 - R) \tag{1}$$

where R is the amount of reduction imparted per direction which is generally assumed to be the same along all the three orthogonal directions, and N is the number of MAF cycles.

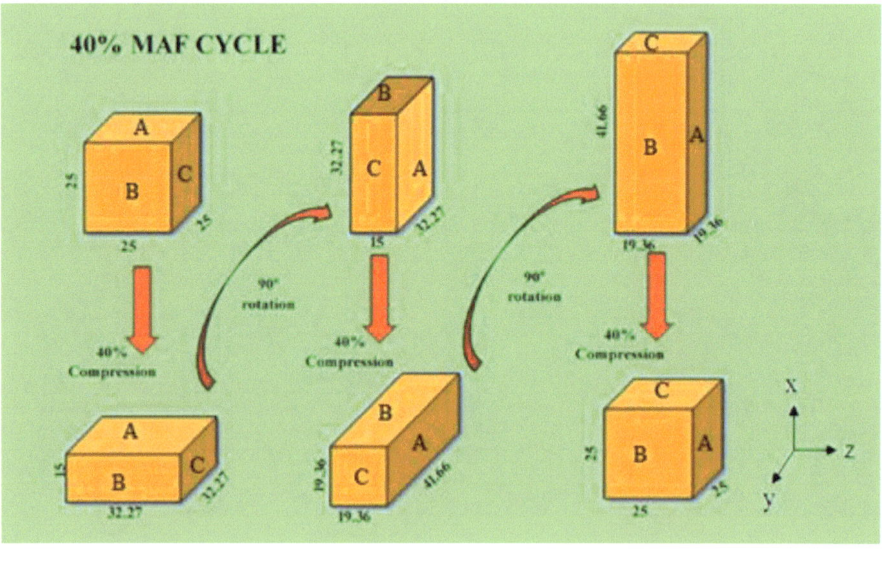

Figure 4.
Typical multi-axial forging cycle for imparting 40% deformation along all the three directions (all dimensions in mm).

Two numbers of cubic specimens with dimension of 2.5 x 2.5 x 2.5 cm^3 were prepared from the water-quenched and air-cooled Ti-6Al-4V material respectively for carrying out MAF experiments at 900°C with a constant true strain rate of 10^{-3}/s. The first sample from the water-quenched and air-cooled material was subjected to 25% reduction along all the three directions while the second sample was imparted with 40% reduction which corresponds to a cumulative strain of 87% for the sample subjected to 25% reduction and 153% for the 40% reduction sample. While carrying of MAF experiment, the sample was first deformed to the required amount in a particular direction (e.g. X direction), the furnace was opened, and then the sample was rotated and positioned for deforming along the next direction (e.g. Y direction). The furnace was closed and sample was reheated to the desired temperature and held at this temperature for 30 min before imparting the desired amount of deformation along this direction (e.g. Y direction). The procedure was repeated for deforming along the third direction (e.g. Z direction). After completion of desired number of MAF cycle, i.e. after imparting desired amount of reduction along all the three directions as shown in **Figure 4**, the samples were water-quenched to free the microstructure of the material.

Microstructure of Ti-6Al-4V with an initial lamellar structure subjected to a deformation of 25% and 40% along all the three orthogonal direction using MAF process is shown in **Figure 5a** and **b** respectively. No major microstructural change is observed in the material subject to 25% when compared with the initial starting microstructure expect for coarsening of α lamellae and few isolated regions of α globularisation within the grain and at grain boundaries. Partial globularisation of α can be observed at all the three faces of the same subjected to 40% deformation through MAF. Though isolated globularisation of grain boundary α was observed, the prior β grains were found to be intact and distinct.

The microstructures of Ti-6Al-4V with martensitic structure subjected to 25% and 40% reduction using MAF process are shown in **Figure 6a** and **b** respectively. Similar to that of the air-cooled structure, martensitic structure also shows increasing

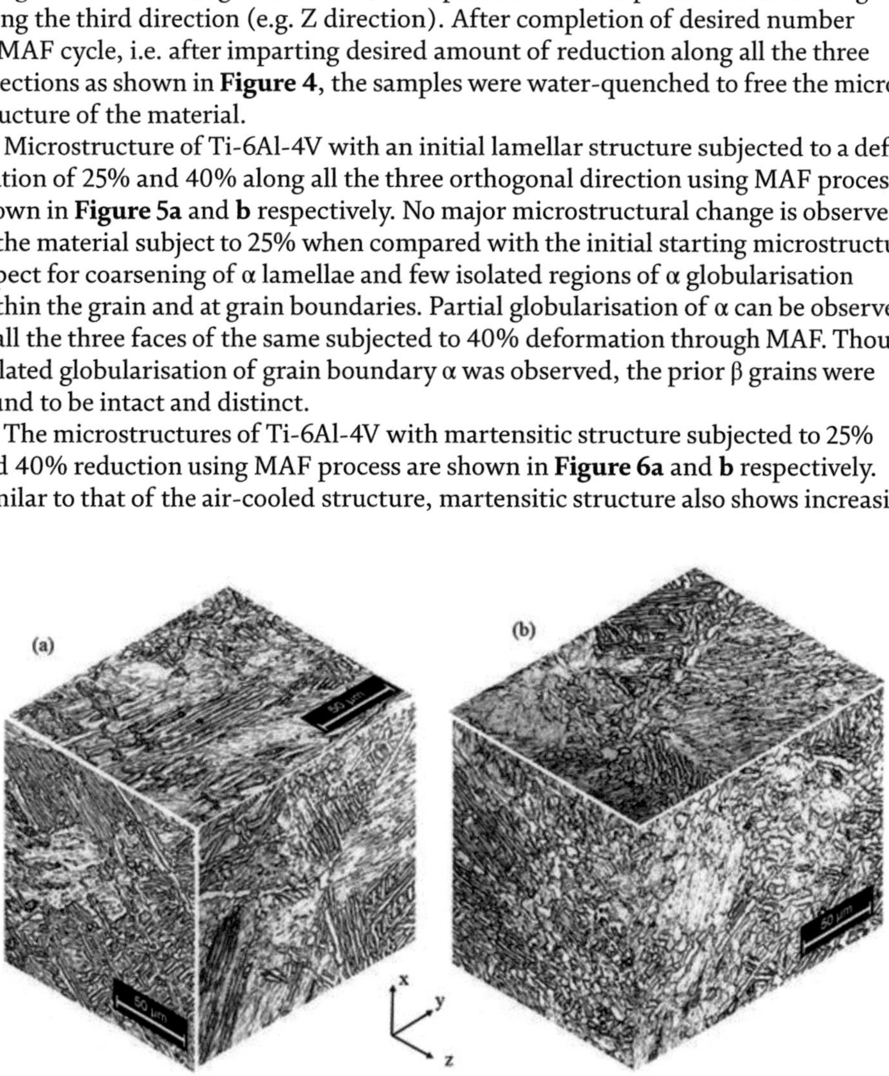

Figure 5.
Microstructure of Ti-6Al-4V with lamellar microstructure subjected to (a) 25% and (b) 40% deformation along all the three orthogonal directions through MAF at 900°C with a strain rate of 10^{-3}/s.

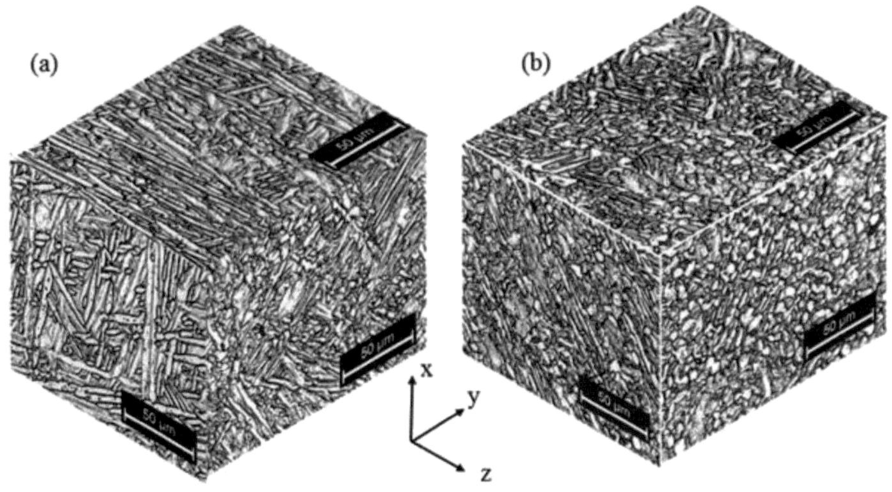

Figure 6.
Microstructure of Ti-6Al-4V with martensitic microstructure subjected to (a) 25 and (b) 40% deformation per direction through MAF process at 900°C with a strain rate of 10^{-3}/s.

globularisation with increasing amount of deformation. However, the breaking of the prior β grain boundaries and globularisation can be observed to be more in martensitic structure when compared with the lamellar structure.

3.3 Globularisation fraction

The globular α volume fraction was estimated in Ti-6Al-4V with martensitic and lamellar starting structure subjected to uniaxial (monotonic, i.e. no change in strain path) and multi axial (non-monotonic, i.e. changing strain path as the sample is rotated) deformation and is shown in **Figure 7**. It can be seen that, irrespective of the deformation process whether it is monotonic (uniaxial) or non-monotonic (multi-axial), the volume fraction of globularised α increases with increasing strain. The rate of increase in globularised α volume fraction is observed to be high for Ti-6Al-4V with martensitic structure when compared with the material with lamellar structure. This observation here concurs with the report of Shell et al. [18] where they have attributed the increased kinetics of globularisation to α lamellae thickness. When the α lamellae is thin, it is relatively easier for the β phase to penetrate the lamellae and transverse boundary when compared with thick lamellae. It can be noted that for an equivalent strain of 0.87 which corresponds to 58% reduction through uniaxial or monotonic deformation and 25% deformation along all the three directions through multi-axial or non-monotonic deformation, altering the strain path through multi-axial deformation results in reduced globularisation. However, with increasing equivalent strain from 0.87 to 1.51, i.e. increasing the deformation from 25 to 40% per direction in multi-axial deformation, the fraction of α phase globularised is quite comparable to that obtained through uniaxial deformation (with 78% reduction) as shown in **Figure 7**.

It was reported by Banerjee et al. [19] that when Ti alloys are subjected to thermomechanical processing in the α/β regime, an equiaxed microstructure can

DOI: http://dx.doi.org/10.5772/intechopen.108463

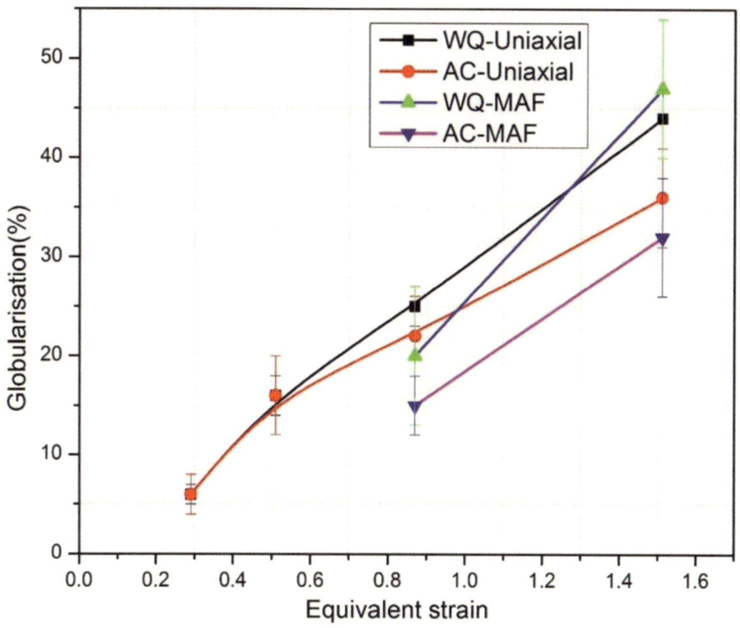

Figure 7.
Effect of strain and strain path on globularisation of α lamellae in Ti-6Al-4V.

be obtained only if the amount of deformation is greater than 30%. As stated earlier, the CRDX or globularisation of α lamellae depends on the formation of α/α interface by shearing, subgrain, buckling and kinking, etc. and the penetration of β phase. As the formation of α/α interface is a result of two contending processes of dislocation accumulation and annihilation, it is proposed that altering strain path before attaining a limiting critical strain leads to large annihilation of dislocations and a possible reduction or disappearance of the substructure. In the material subjected to 25% deformation through MAF, the strain path is changed before reaching the critical amount of deformation (~30%), and the fraction of globularisation is less due to large annihilation of dislocations. With increasing strain from 87 to 151%, adequate strain is available for accumulation of dislocations and formation of stable substructure. Hence, altering the strain path does not influence the fraction of globularisation. From **Figure 7**, it can be readily inferred that 40% deformation through non-monotonic MAF does not cause significant reduction in the globularisation kinetics when compared with uniaxial deformation. Concomitantly, no significant improvement has been observed as reported by Salischev et al. [4].

On the basis of above study, it can be readily inferred that a martensitic structure exhibits better globularisation kinetics in comparison to lamellar structure. Altering the strain path through non-monotonic multi-axial deformation below a critical limiting strain results in reduced globularisation, whereas beyond the critical strain, the results are quite comparable. The critical deformation limit has been identified to be ≥40% through the current investigation. Though required globularisation can be achieved by uniaxial deformation, it results in altering the dimension of the work piece, whereas multi-axial deformation process which does not cause any concomitant change in the external dimension [20]. Further, unlike uniaxial deformation which produces a dead metal zone at the fiction hill, no such un-deformed region is observed in multi-axial deformation; hence, multi-axial deformation process with 40% reduction per direction is best suited to globularise the α lamellae and refine the microstructure to obtain superplasticity.

4. Thermomechanical processing scheme to achieve ultrafine grain structure

To obtain grain refinement and achieve superplasticity, Ti-6Al-4V with a martensitic starting microstructure was subjected to three cycles of multi-axial deformation with 40% reduction per direction by progressively decreasing the deformation temperature at each cycle from 850, 800 and 750°C, i.e. first cycle of MAF is carried out at 850°C, second cycle of MAF on the sample is carried out at 800°C and so on. Microstructures of Ti-6Al-4V obtained after two and three cycles of multi-axial deformation are shown in **Figure 8a** and **b** respectively. It can be seen that three cycles of multi-axial deformation have resulted in complete globularisation of α lamellae.

The mechanism of globularisation or CDRX of α lamellae depends on formation of α/α boundaries which is followed by penetration of β phase into α lamellae to separate the boundary [15]. Formation of stable substructure (α/α boundaries) or low-angle grain boundaries (LAGB) depends on interaction and multiplication of dislocation in the lamellar structure. When the HCP α lamellae c-axis is oriented parallel to the deformation direction, high shear stresses are required for the operation of basal and prism slip which is the major source of slip in HCP structure [17]. Such unfavourable orientated lamellae exhibit no major deformation leading to low dislocation density and less α/α boundaries which reduces the globularisation kinetics. However, rotation of specimen by 90° during subsequent steps of deformation during multi-axial deformation leads to activation of prism slip system in such lamellae due to an increase in value of Schmid factor. Activation of prism slip system initiates the process of slip, and with increasing

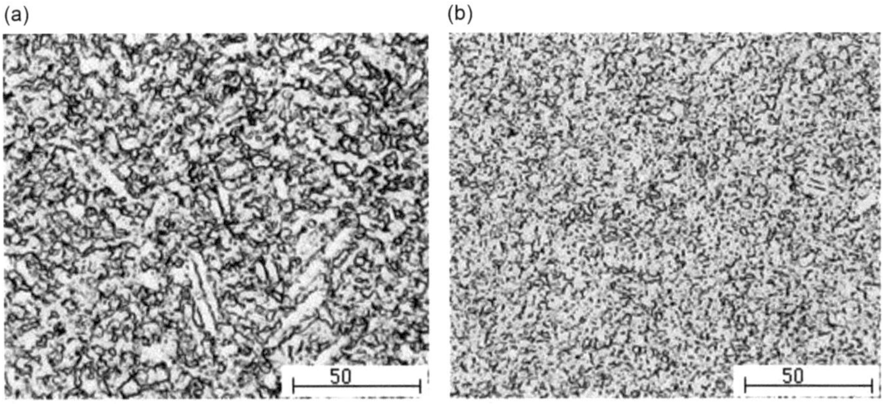

Figure 8.
Microstructure of Ti-6Al-4V after (a) two cycles and (b) three cycles of MAF.

strain and grain rotation, the basal slip system also starts operating leading to formation of substructure, which eventually results in dynamic recrystallised or globularised equiaxed α grain. A comparison of the microstructures shown in **Figure 6b**, **8a** and **b** reveals that with increasing the number of MAF cycle, globularisation increases. It can also be seen that the isolated α lamellae visible in Ti-6Al-4V specimen after two cycles of multi-axial deformation are converted into equiaxed grains after three cycles. Further, prior β grain boundaries were completely destroyed, and they are no longer distinguishable after three MAF cycles.

To gain further insight on the microstructure evolution, the material subjected to three cycles of multi-axial deformation was subjected to electron back scattered diffraction (EBSD) characterisation. The low-angle (LAGB; red line) and high-angle (HAGB; black line) boundary present in the material was estimated to be 24% and 76% respectively. The observed fraction of HAGB and LAGB is consistent with reported literature which states the presence of around 25–30% of LAGB in heavily deformed structures [20]. The average size of HAGB was estimated to be 1.65 ± 0.6 μm. Further, the inverse pole figure (IPF map) shown in **Figure 9(b)** reveals a random texture with no preferred orientation.

Superplasticity occurs by grain boundary sliding accommodated by diffusion at grain boundaries and lattice [21, 22]. A material with high fraction of HAGB will have higher degree of disorder. As diffusion occurs down the potential gradient, large fraction of HAGB will lead to higher gradient which increases the diffusion rate. It has been reported in the literature that the presence of high fraction of HAGB with maximum fraction lying in the range of 30–60° is necessary for ease of grain boundary sliding [4]. Further, a fairly uniform grain structure with grains of similar size is essential for superplastic forming. The presence of large grains and small grains (mixed or bimodal grains) in same microstructure has a negative effect on superplasticity. It has been reported that kinetics of diffusion is very slow around the large grain in comparison to the kinetics around smaller grain structure [23]. In the absence of slow diffusion around large grains, chances of cavitation are higher around the larger grains leading to early failure without appreciable tensile elongation. It is also well known that higher volume fraction

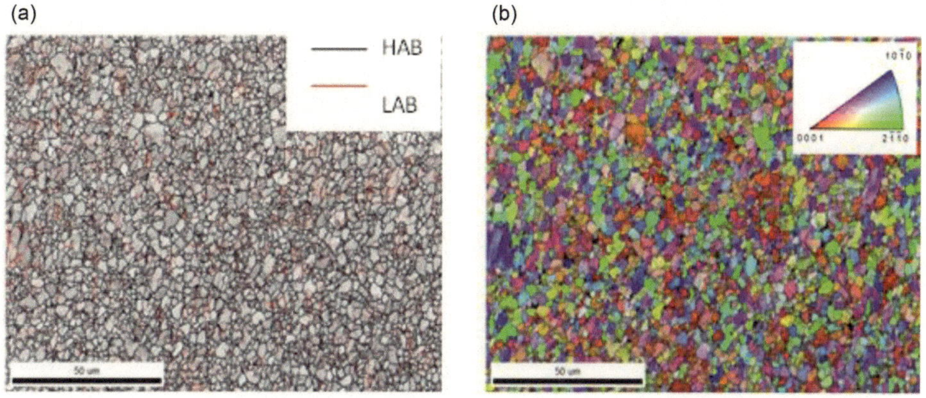

Figure 9.
Microstructure of Ti-6Al-4V after three cycles of MAF(a) band contrast image highlighting HAGB (black line) & LAGB (red line) (b)I PF colour map obtained through EBSD.

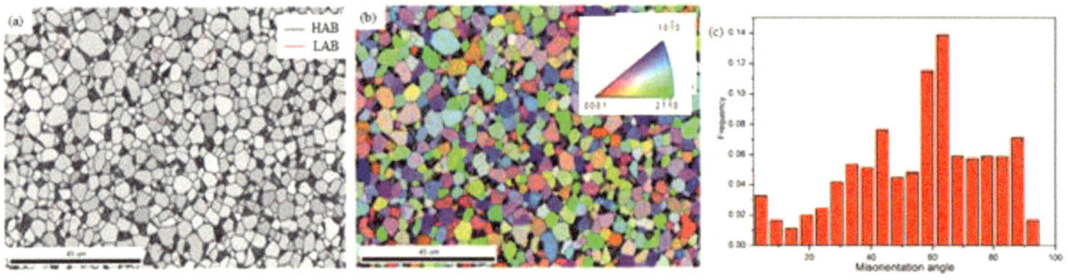

Figure 10.
Microstructure of Ti-6Al-4V after three cycles of MAF and heat treatment at 850°C for 2 h followed by air cooling (a) band contrast image highlighting HAGB (black line) & LAGB (red line) (b) IPF colour map (c) misorientation plot obtained through EBSD.

of second phase improves superplastic property due to easier α/β grain boundary sliding [24]. Further, in titanium alloys, the diffusivity in BCC β grain is about of two orders of magnitude higher than α alloys [1, 2, 20]. Higher diffusivity of β phase improves superplastic property of titanium alloy. However, at the same time, β phase grows much faster than α phase, and hence, α phase helps in pinning the grain boundary of β and does not allow the grains to grow. Ideally, presence of both phases is essential for easier grain sliding and achieving optimal superplastic property [25].

As the material after three cycles of multi-axial deformation exhibits nearly 24% low-angle boundaries and predominantly α phase, it is essential to convert this microstructure into a one that can exhibit superplasticity. To improve microstructure and to obtain uniform grain size distribution with increased volume fraction of β phase and higher fraction of high-angle grain boundary, annealing of the deformed material was carried out at 850°C for 2 h followed by air cooling. A higher temperature was used for annealing to obtain the desired α and β proportions in the material. From the EBSD band contrast image shown in **Figure 10a**, the β phase present in the heat treated material was estimated to be ~15%. The fraction of HAGB in the heat-treated material was estimated to be 95% with an average grain size of 2.53 ± 0.65 μm. The IPF map shown in **Figure 10b** clearly indicates the random orientation of the grains. A random texture is important for a structure to exhibit superplastic behaviour since cavitation may occur along the transverse direction during deformation due to strain incompatibility [21]. The misorientation profile of the grains shows significant improvement in HAB with a high fraction of grains boundary in the range of 30–60°. The resultant microstructure satisfies the condition of superplasticity as a major mode of deformation.

5. Superplastic domain identification and validation

In order to identify the temperature-strain rate domain under which the material exhibits superplasticity, the SRS of the material subjected to three cycles of MAF and heat treatment was evaluated by carrying out isothermal hot compression tests over a range of temperature and constant true strain rate. Using the flow curves, the SRS of the material was estimated using standard relations [26] and plotted as a function of temperature and strain rate as shown in **Figure 11**. It can be seen

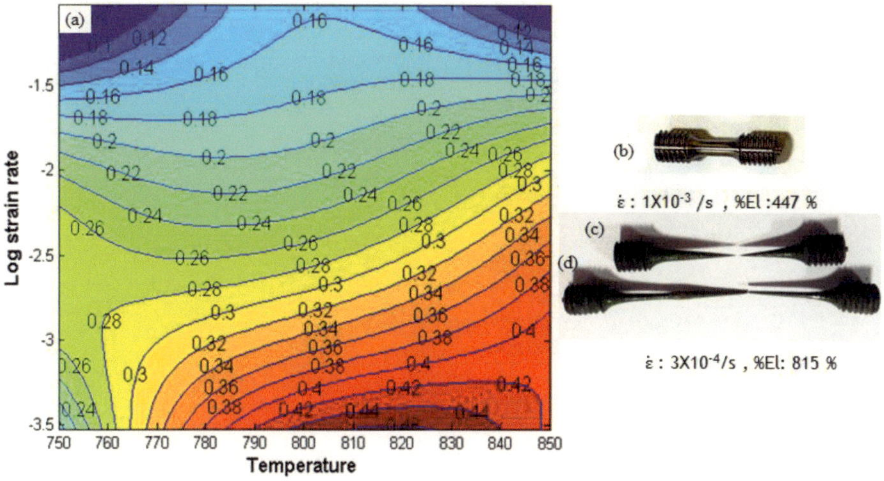

Figure 11.
(a) Strain rate sensitivity map at ε = 0.5 (b) initial tensile specimen (gage length = 6 mm, diameter = 4 mm) (c) sample after tensile deformation at 820°C with a strain rate of 10^{-3}/s strain rate and (d) sample after tensile deformation at 820°C with a strain rate of 3×10^{-4}/s strain rate.

that the material exhibits a maximum SRS of 0.46 between 810°C and 825°C for a constant strain rate of 3×10^{-4}/s.

In order to validate the domain identified, cylindrical tensile samples of 0.6 cm gage length and 0.4 cm diameter (**Figure 11b**) were prepared from the material subjected to three cycles of MAF and heat treatment. Tensile tests were carried out at 820°C with a strain rate of 10^{-3} and 3×10^{-4}/s. A maximum elongation of 815% was obtained in the sample tested with a strain rate of 3×10^{-4}/s (**Figure 11d**). As the strain rate is increased to 10^{-3}/s, the % elongation obtained decreases from 815 to 447% as shown in **Figure 11c**. During tensile deformation, necking starts in the weak part of the structure and further deformation gets concentrated at this necked region. A tri-axial state-of-stress exists in the neck region and strain rate in this region does not follow the strain rate of the specimen. The strain rate in the region depends on the rate of decrease of the area which is given by [26]:

$$\frac{dA}{dt}=\left(\frac{P}{C}\right)^{\frac{1}{m}}\left(\frac{1}{A^{(1-m)/m}}\right) \tag{2}$$

As the rate of decrease in area is inversely proportional to strain rate sensitivity (m), a higher m value leads to slower strain rate in the necked region. So a higher elongation is observed during tensile deformation.

In order to further substantiate the findings, EBSD characterisation of the material subjected to compression test (after three cycles of multi-axial deformation and heat treatment) at 800°C with a strain rate of 3×10^{-4}/s was carried out to evaluate the grain size, boundary fractions, etc. From the band contrast image shown in **Figure 12**, the fraction of HAB and the average grain size were estimated to be 83% and 2.64 ± 0.89 μm respectively. Comparing the microstructure

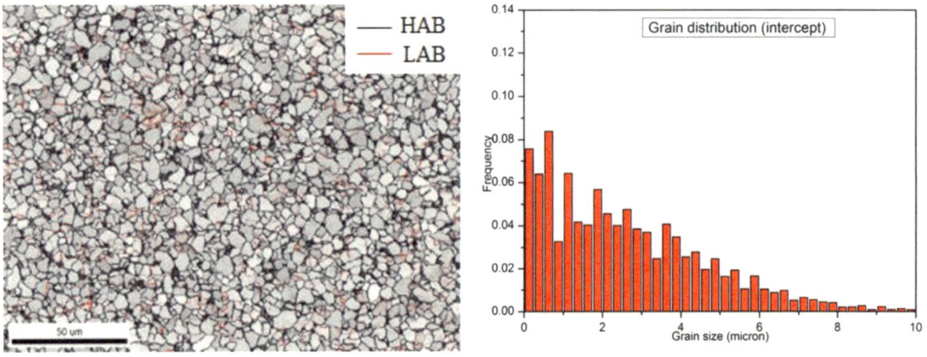

Figure 12.
Band contrast image of fine grained Ti-6Al-4V after hot compression at 800°C and strain rate of 0.0003 S^{-1} highlighted with HAGBs (black), LAGBs (red).

obtained in the material after deformation in the superplasticity domain and that obtained after three cycles of multi-axial deformation and heat treatment (**Figure 10a**) clearly indicates no major change in grain size. Therefore, from the observations on % elongation and microstructural features, it can be confirmed that the heat-treated material after subjecting to three cycles of multi-axial deformation exhibits superplasticity between 800 and 840°C when deformed with a strain rate of 3×10^{-4}/s

6. Scaling-up

In order to ensure viability of the identified thermomechanical scheme for industrial scale processing, a large-size billet of 15 cm × 12 cm × 9 cm was prepared and subjected to MAF using 2000MT hydraulic forge press. **Figure 13** compares the size of scaled up Ti-6Al-4V billet with that of smaller-size specimen. The microstructure evolution in the large billet after each cycle of MAF is shown in **Figure 14**. It can be clearly seen that the globularisation fraction increases with increasing cycle and a completely globularised structure is achieved after three cycles of MAF. Post-deformation heat treatment for the material was carried out at 850 and 900°C for 2 h followed by air cooling, and the resulting microstructures are shown in **Figure 15**. Heat treatment led to increase in the volume fraction of beta phase with slight coarsening in the grain size.

In order to validate super plasticity, standard cylindrical tensile samples of 2.0 cm gage length and 0.4 cm diameter were prepared and subjected to tensile testing. Tensile tests were carried out at 820°C with a Strain rate of 3 x 10^{-4} and 10^{-3}/s. Tensile test was carried out for as deformed (three cycles of MAF) specimen also. It can be seen from **Figure 16** that a maximum elongation of 640% has been obtained for the material.

The result is comparable with the one achieved during tensile test of smaller-size MAF processed specimen. The obtained results have been also compared with available literature as shown in **Table 1**. It can be seen that obtained m values and elongation values are comparable with the data available in literature. Elongation obtained during tensile test of specimen extracted from smaller-size MAF specimen and large-size MAF billets clearly indicates superplastic behaviour of three cycles of MAF

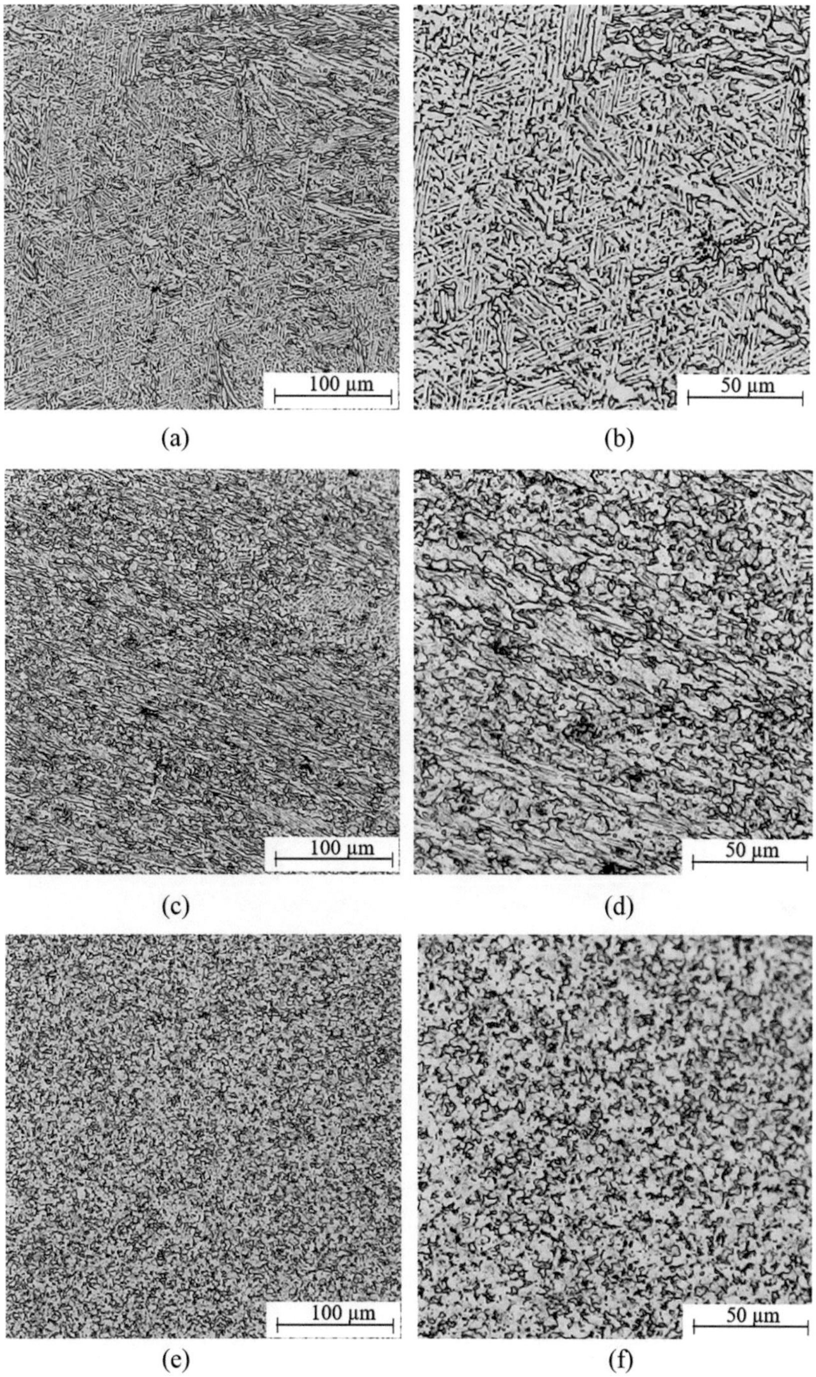

Figure 13.
Typical microstructure of large size Ti-6Al-4V after (a,d) first MAF cycle (b,e) second MAF cycle (c, f) third MAF cycle.

Ti-6Al-4V alloy. Hence, it can be concluded that the TMP scheme established using smaller sample is repeatable and is validated by repeating the process using 2000 T hydraulic Forge press.

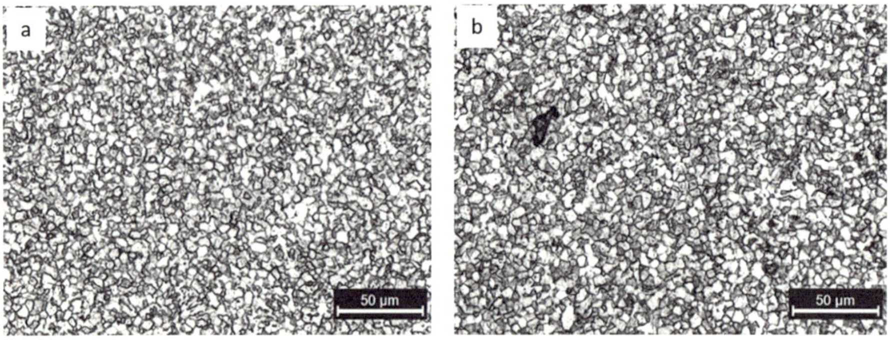

Figure 14.
Microstructure of Ti-6Al-4V obtained after third MAF cycle followed by heat treatment at (a) 850°C/2 h/ac and (b) 900°C/2 h/ac.

Figure 15.
Comparison of scaled-up Ti-6Al-4V billet and small size specimen.

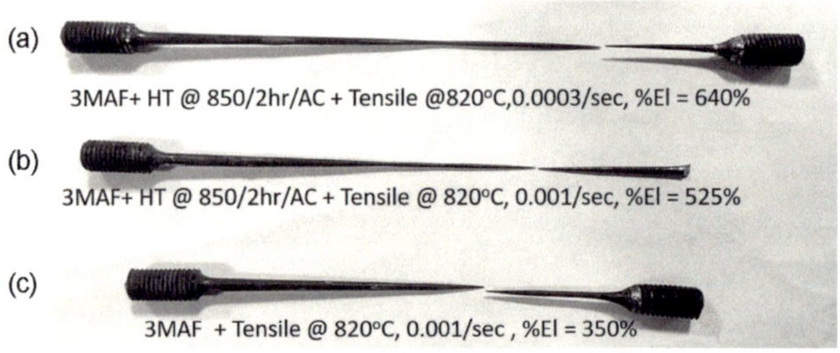

Figure 16.
Tensile elongation of Ti-6Al-4V subjected to three cycles of MAF followed by heat treatment at 850°C/2 h/ac and (a) tested at $\acute{\varepsilon}$ = 0.0003/s, T = 820°C (b) tested at $\acute{\varepsilon}$ = 0.001/s, T = 820°C (c) Tensile elongation of Ti-6Al-4V subjected to three cycles of MAF tested at $\acute{\varepsilon}$ = 0.001/s, T = 820°C.

Processing condition	Grain size (μm)	Temp (°C)	Strain rate (s^{-1})	Elongation (%)	m	Ref
Rolled sheet	0.3	775	1×10^{-4}	800	0.45	[27]
		775	1×10^{-3}	600	0.45	
		875	1×10^{-4}	720	0.41	
		875	1×10^{-3}	380	0.41	
	3	775	1×10^{-4}	200	0.32	
		875	1×10^{-4}	420	0.33	
		875	1×10^{-3}	390	0.33	
SPD processed bulk samples	0.2	600	5×10^{-4}	500	0.34	[28]
	5	800	5×10^{-4}	600	0.4	
Hot Rolled	3	700	1×10^{-4}	583	0.66	[29]
SPD processed (HPT)	0.2	650	1×10^{-4}	568	0.36	[30]
		750	1×10^{-2}	504	0.46	
DMRL 3 cycle MAF + Heat treated	2.53	820	3×10^{-4}	640	0.46	
		820	1×10^{-3}	525	0.38	

Table 1.
Comparison of achieved superplastic property with available literature.

7. Conclusions

From the study it can be concluded that three cycles of MAF with 40% reduction per direction (cumulative effective strain of ~4.6) leads to complete globularisation of martensitic structure with an average α grain size of 1.65 ± 0.6 μm. Annealing of the deformed material at 850°C increases HAB fraction and β phase volume fraction but with a marginal increase in grain size (2.53 ± 0.65 μm). The MAF + annealed material exhibits a maximum SRS of 0.46 when deformed between 810°C and 825°C with constant strain rate of 3×10^{-4}/s. A maximum tensile elongation of 815% was obtained with strain rate of 3×10^{-4}/s at 820°C. TMP designed was implemented on a large-size work piece under near isothermal condition, and the process was found to be reliable and repeatable to obtain superplasticity in bulk Ti-6Al-4V in large-size specimen for aeroengine applications.

Acknowledgements

The funding provided by Defence Research and Development Organisation is acknowledged.

Conflict of interest

The authors declare no conflict of interest.

Author details

Bipin Kedia and Ilangovan Balasundar*
Near Net Shape Group, Directorate of Advanced Materials and Manufacturing Processes, Defence Metallurgical Research Laboratory, Hyderabad, India

*Address all correspondence to: i-balasundar.dmrl@gov.in

References

[1] Lutjering G, Williams JC. Titanium. Berlin (Deutschland): Springer; 2003

[2] Leyens C, Peters M. Titanium and Titanium Alloys. Chichester (England): John Wiley & Sons Inc; 2003

[3] Weiss I, Froes FH, Eylon D, Welsch GE. Modification of alpha morphology in Ti-6Al-4V by Thermo-mechanical processing. Metallurgical and Materials Transactions A. 1986;**17**:1935-1947

[4] Salishchev GA, Mironov YS, Zherebtsov SV. Mechanism of sub- microcrystalline structure formation in titanium and two – Phase titanium alloy during warm severe processing. Reviews on Advanced Materials Science. 2006;**11**:152-158

[5] Zhang T, Liu Y, Sanders DG, Liu B, Zhang W, Zhou C. Development of fine grained titanium 6Al-4V alloy sheet material for low temperature superplastic forming. Material Science and Engineering A. 2014;**608**:265-272

[6] Salishchev GA, Galeyev RM, Valiakhmetov OR, Safiullin RV, Lutfullin RY, Senkovv ON, et al. Highly superplastic Ti-6AI-4V sheet for superplastic forming and diffusion bonding. Materials Technology. 2000;**15**(2):133-135

[7] Maoheny MW. Technical note 5A – Superplastic forming of titanium alloys. In: Boyer R, Welsch G, Collings EW, editors. Material Properties Handbook. Metals Park, Ohio, USA: ASM International; 1994. pp. 1101-1109

[8] Zherebtsov SV, Salishchev AG, Galeyev RM, Valiakhmetov OR, Mironov Y, Semiatin SL. Production of sub microcrystalline structure in large-scale Ti–6Al–4V billet by warm severe deformation processing. Scripta Materialia. 2004;**51**:1147-1151

[9] Poths RM, Angella G, Waynne BP, Rainforth WM, Semiatin SL, Beynon JH. Effect of strain reversal on the dynamic spherodisation of Ti-6Al-4V during hot deformation. Metallurgical and Materials Transactions A. 2004;**35A**:2993-3001

[10] Vander Voort GE. Metallographic Principles and Practices. New York: McGraw-Hill; 1984

[11] Balasundar I, Raghu T, Kashyap BP. Correlation between microstructural features and creep strain in a near-α titanium alloy processed in the α+β regime. Materials Science and Engineering A. 2014;**609**:241-249

[12] Boyer R, Welsch G, Coolings EW. Material Properties Handbook Titanium Alloys. Ohio: ASM International; 1996

[13] Prasad YVRK, Seshachayulu T. Modelling of hot deformation for microstructural control. International Materials Reviews. 1998;**43**:243-258

[14] Humphreys FJ, Hartherly M. Recrystallization and Related Annealing Phenomena. II ed. Oxford: Elsevier; 2004

[15] Margolin H, Cohen P. Evolution of Equiaxed Alpha Morphology of Phases in Ti-6Al-4V Alloy. Kyoto, Japan: Titanium 80- Science and Technology; 1980. pp. 1555-1561

[16] Balasundar I. Modeling the High Temperature Flow Behaviour and Study of Structure-Property Correlation in near-α Titanium Alloy [Ph.D thesis]. Bombay, India: Indian Institute of Technology; 2013

[17] Bieler TR, Semiatin SL. The origins of heterogeneous deformation during primary hot working of Ti-6Al-4V. International Journal of Plasticity. 2002;**18**:1165-1189

[18] Shell EB, Semaitin SL. Effect of initial microstructure on plastic flow and dynamic Globularisation during hot working of Ti-6Al-4V. Metallurgical and Materials Transactions A. 1999;**30**:3219-3229

[19] Banerjee D, Krishnan RV. Challenges in alloy design: Titanium for the aerospace industry. Proceedings of Indian Academic Science. 1981;**4**:21-39

[20] Kaibyshev OA, Utyashev FZ. Superplasticity: Microstructural Refinement and Superplastic Roll Forming. Vol. 3. Virginia, USA: ISTC Science and Technology Series; 2005

[21] Nieh TG, Wadsworth J. Superplasticity in Metals and Ceramics. 1st ed. Cambridge: Cambridge University Press; 1987. pp. 79-80

[22] Langdon TG. An evaluation of the strain contributed in superplasticity. Material Science and Engineering A. 1994;**174**:225-230

[23] Paton NE, Hamilton CH. Microstructural influence on superplasticity in TI-6Al-4V. Materials and Metallurgical Transactions A. 1979;**10**:241-250

[24] Sieniawski J, Motyka M. Superplasticity in titanium alloys. Journal of Achievements in Materials and Manufacturing Engineering. 2007;**24**:123-130

[25] Kim JS, Kim JH, Lee YT, Park CG, Lee CS. Microstructural analysis on boundary sliding and its accommodation mode during superplastic deformation of Ti–6Al–4V alloy. Materials Science and Engineering A. 1999;**263**:272-280

[26] Dieter GE. Mechanical Metallurgy. SI Metric edition. London: McGraw-Hill Book Co.; pp. 293-300

[27] Patnakar SN, Escobedo JP, Field DP, Salishchev G, Galeyev RM, Valiakhmetov OR, et al. Superior superplastic behavior in fine grained Ti-6Al-4V sheet. Journal of Alloys and Compounds. 2002;**345**:221-227

[28] Salishchev GA, Valiakhmetov OR, Valitov VA, Mukhtarov SK. Sub-microcrystalline and nanocrystalline structure formation in materials and search for outstanding superplastic properties. Material Science Forum. 1994;**170-172**:121-130

[29] Matsumoto H, Nishihara T, Velay V, Vidal V. Superplastic property of the Ti-6Al-4V alloy with ultrafine grained heterogenous microstructure. Advanced Engineering Materials. 2017;**00**:1700317

[30] Sergueeva AV, Stolyarov VV, Valiev RZ, Mukherjee AK. Superplastic behaviour of ultrafine grained Ti-6Al-4V alloys. Materials Science and Engineering. 2002;**A323**:318-325

Section 2

Extraction

Chapter 3

Titanium Extraction Metallurgy Developments and Control of Impurity Elements

Chaganti R.V.S. Nagesh

Abstract

Titanium extraction metallurgy poses numerous challenges owing to a combination of various characteristics such as high chemical reactivity, high melting point, strong affinity towards oxygen and nitrogen, pyrophoricity of nascent sponge. Kroll process of magnesium reduction of has become the widely employed titanium sponge production technology. Sodium reduction of $TiCl_4$ known as Hunter's process has also been employed for the industrial production of titanium sponges for about two decades. Subsequently quoting techno-economic reasons, the Hunter sponge plants across the world have been closed. There have been several efforts over the years to evolve an alternative to the Kroll process mainly to achieve a simple and cost and energy-effective titanium extraction process. Control of impurity elements in the titanium metal during the metal extraction process assumes greater importance as thermodynamics does not favour any purification method to be employed for the metal. This chapter brings out historical developments in titanium extraction metallurgy and highlights recent developments as well to produce high-purity titanium sponges required for titanium alloys for the end applications across various sectors.

Keywords: titanium, metal extraction, purity, quality control

1. Introduction

Titanium is an element with atomic number 22 and placed in the Group IV B of the periodic table. Discovered by an English chemist, William Gregor in 1791, assumed its name 'titanium' when a German chemist, M.H. Klaproth found it in 1795 as a new element in the mineral rutile and called it 'titanium' (titans is the goddess of earth in Greek). Titanium possesses a melting point of 1663°C and a density of 4.5 g/cm^3. Lightweight, high specific strength coupled with excellent corrosion resistance resulted in titanium emerging as the structural material of choice for a variety of applications in aerospace and many chemical industries. The titanium also has special properties of body compatibility, non-magnetism and non-toxicity. Titanium has different oxidation states and is highly reactive and forms alloys and compounds with a large number of elements. It is because of the outer thin oxide film, the metal exhibits superior corrosion resistance to a wide range of chemicals, seawater and aggressive media. The ability to form different types of attractive/lustrous shades of colours on the surface on anodizing led to its use in artificial jewellery and various types of eye-catching consumer durables.

Titanium exhibits allotropy and changes its crystal structure from hcp to bcc on heating at 882°C. Important properties of titanium of interest for a material scientist/metallurgist include its amenability to heat treatment to achieve tailor-made properties for different end applications. A large number of titanium alloys of type alpha, alpha plus beta, beta and gamma phases are developed and put into use for various types of applications up to a temperature of 550°C.

Titanium is plentifully available in nature. In the early days, it was referred to as 'rare metal' mainly because of its limited usage on account of its expensive nature. Constituting about 0.63% of the earth's crust, titanium is the ninth most abundant element and fourth most abundant structural metal (after iron, magnesium and aluminium). The two most important and widely available titanium minerals are ilmenite ($FeOTiO_2$) and rutile (TiO_2). Ilmenite is subjected to chemical beneficiation or thermal smelting to prepare synthetic rutile or titania slag, which contains increased TiO_2 content.

2. Emergence of titanium extraction metallurgy

Extractive metallurgy of titanium is complicated due to the high melting point, high chemical reactivity and pyrophoric nature (catches fire under frictional forces) of nascent titanium metal. Vacuum or inter-gas atmosphere is essential for producing titanium metal. Historically preparation of pure titanium metal from the oxide concentrate has been a challenging task. Early efforts to prepare the metal directly from its oxide had been futile because of the large amount of oxygen retained in the product. Oxygen, nitrogen, carbon and iron are the most important impurity elements in titanium, which adversely affect the ductility and mechanical properties of the metal. Realizing the implications of preparing the metal directly from the oxide, early efforts shifted the focus to attempt metal separation from a non-oxygen-bearing compound. Thus, chloride metallurgy came into existence where a metal chloride is prepared by high-temperature chlorination of metal oxide and subsequently the metal chloride is reduced/electrolysed to prepare the metal.

In 1910, Hunter developed sodium reduction of titanium tetrachloride ($TiCl_4$) to prepare high-purity titanium metal. The process was later on extensively studied and developed by Imperial Chemical Industries, UK, for industrial implementation. Some industries based on this process in UK and Japan had been operating till the early 90s. In 1925, van Arkel and de Boer developed the process of dissociation of titanium iodide (TiI_4) on a tungsten filament to produce high-purity titanium metal. In 1937, W J Kroll worked on magnesium reduction of $TiCl_4$ to produce high-purity metal and patented the process. The Kroll process was subsequently developed by the US Bureau of Mines for industrial-scale implementation. Simultaneously in the former USSR and Japan also similar developments took place. Fused salt electrolysis of $TiCl_4$ to prepare titanium was extensively studied by Dow Howmet USA, Reactive Metals Inc., UK, and Electrochemica Marco Ginatta, Italy, and pilot scale plants had been operating based on this process till recently. These three processes *viz*. Hunter process of sodium reduction, Kroll process of magnesium reduction and fused salt electrolysis of $TiCl_4$ are proven and are considered to be established methods of titanium sponge production. In all three processes, nascent metal formed is in the form of a porous aggregate of titanium metal particulates and termed as 'titanium sponge'. Titanium sponge is the basic raw material for the manufacture of titanium/titanium alloys, which takes place through vacuum arc melting of sponge into ingot followed by conventional metal working techniques to obtain the desired product for end use. A detailed description of the processes along with historical developments in metal extraction is well documented in the literature [1–3]. Some of the experiences gained in the Kroll titanium sponge technology development program [4] successfully conducted at the Defence Metallurgical Research Laboratory, Hyderabad, India, are shared in this presentation.

Subsequently, there have been several efforts the world over to develop a simple, energy and cost-effective and environmentally friendly titanium extraction process. Most of these efforts focus to cut down the process steps, evolving a continuous process and/or directly obtaining titanium from TiO_2. The majority of these advanced processes result in titanium powder, which can be utilized in preparing the desired products through additive manufacturing or 3D printing.

3. Established methods of titanium sponge production

As discussed, the preparation of $TiCl_4$ is an essential step in the production of titanium sponges. $TiCl_4$ is mainly prepared by high-temperature carbo-chlorination of titanium mineral concentrate (rutile/beneficiated ilmenite) in a fluidized bed-type refractory lined furnace at a temperature of about 1000°C. The tetrachloride obtained on chlorination is relatively impure containing chlorides of other metals such as Fe, Sn, Si, Al, V. $TiCl_4$ is an aggressive chemical and readily reacts with atmospheric moisture and hydrolyzes forming thick fumes of HCl and oxy-chloride. Hence, handling and purification of $TiCl_4$ is hazardous and to be carried out cautiously. Before using for the production of titanium sponge, $TiCl_4$ is to be purified sufficiently, which is generally done by employing fractional distillation and precipitation techniques.

Thus, the established methods of titanium sponge production essentially involve the process steps (**Figure 1**) of chlorination of beneficiated ilmenite or rutile in the presence of carbon at a temperature of about 950C to produce $TiCl_4$, purification of the tetrachloride by fractional distillation to obtain high-purity $TiCl_4$ followed by either sodium reduction (Hunter process) or magnesium (Kroll process). Alternatively, titanium sponge can also be produced by fused salt electrolysis of $TiCl_4$ in LiCl-KCl melt. A brief description of these three processes is provided in the sequel before a discussion is taken up on 'control of impurity elements' in the product.

4. The Hunter process

In this Hunter process, $TiCl_4$ is reduced with sodium metal and the following chemical reaction represents the reduction process:

$$TiCl_4 + 4Na = Ti + 4NaCl,\ \Delta H^{o}_{800C} = -653.7\ \text{kJ/mole}. \quad (1)$$

The reduction process is conducted in an inert gas atmosphere using a steel crucible. $TiCl_4$ is fed into the reaction crucible, which is holding a bed of molten sodium. The temperature of reduction needs to be precisely controlled within the small range

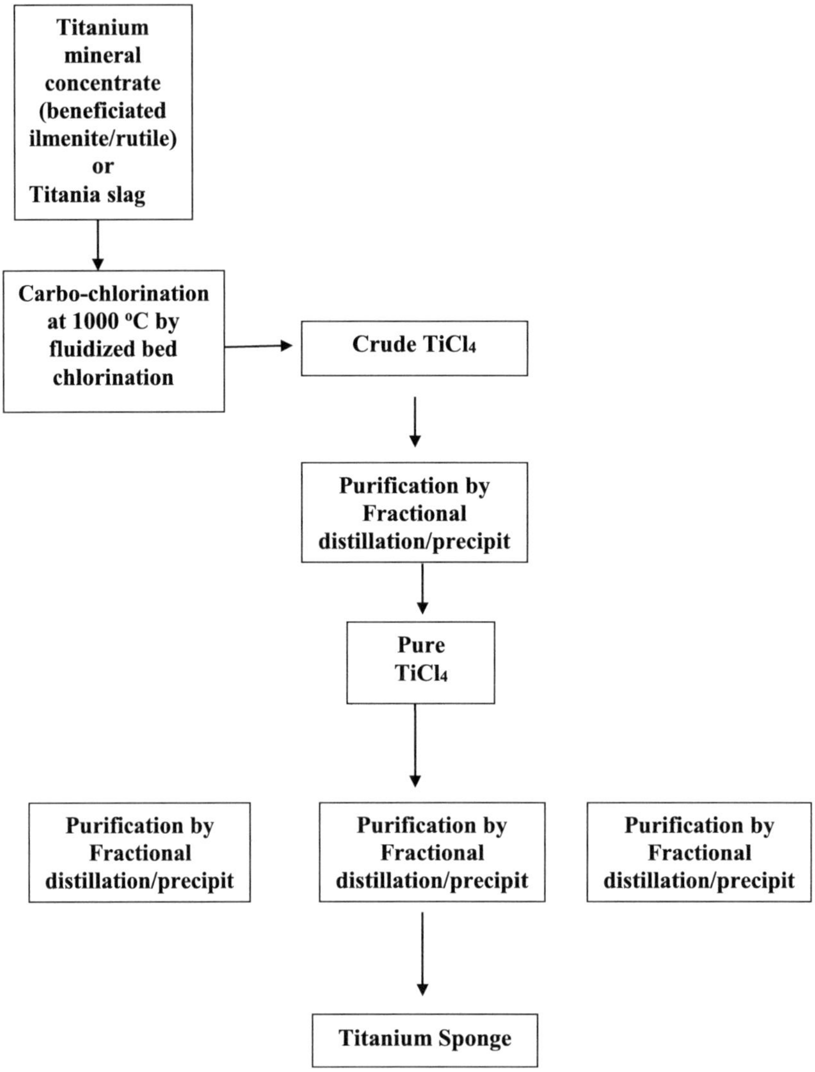

Figure 1.
Process flow diagram for established methods of titanium sponge production.

of melting point of NaCl (801C) and boiling point of sodium metal (887C) for ensuring a smooth reduction process. The main product of the titanium sponge needs to be separated from adhered by-product (NaCl) and leftover reductant (sodium), which is done by water leaching of the reaction product Later on, however, the process underwent several modifications and improvements compared to the method originally employed by Hunter. Nippon Soda, Japan and Deeside Titanium, UK-operated titanium sponge production plants employing this process. However, the process has several setbacks such as very high exothermic heat generation, highly reactive species, very close window of process operating parameters, difficulty in recovering by-product/effluent disposal, etc., in addition to the hazardous nature of handling sodium. Currently, no industry is operating in the world based on this process for titanium sponge production.

5. The Kroll process

The Kroll process of magnesio-thermic reduction of $TiCl_4$ to produce titanium sponge is represented by the following chemical reaction.

$$TiCl_4 + 2Mg = Ti + 2MgCl_2,\ \Delta H^o{}_{800C} = -325\ kJ/mole. \quad (2)$$

The reaction takes place between liquid magnesium and gaseous $TiCl_4$ (boiling point 136C) forming solid titanium. There is a possibility of a large number of reactions involving also lower chlorides of titanium viz. $TiCl_3$ and $TiCl_2$ as presented in **Table 1**. The standard enthalpy and entropy changes of the reactions which indicate the thermodynamic possibility and exothermic/endothermic nature of reactions are presented in the table. The overall reduction reaction is highly exothermic and necessitates external cooling of the reactor for controlling the reaction temperature.

Physico-chemical aspects of reaction chemistry and some aspects of reaction mechanism and titanium sponge formation in a Kroll reduction reactor are discussed in detail in the literature [5, 6]. Also, phase equilibria in the system, Mg-Ti-Cl over a wide temperature range of 500–2000°C, were studied. The extent to which a given reaction would occur is influenced by several factors such as temperature, the physical state of the reactants, mutual solubilities of the substances, surface and interfacial phenomena, reaction kinetics and heat and mass transfer in the system.

From a thermodynamic analysis of various possible reactions involved in the magnesio-thermic reduction of $TiCl_4$, the following inferences were drawn:

- When gaseous $TiCl_4$ reacts with liquid magnesium, reduction of $TiCl_4$ to $TiCl_2$ is the most probable.

- When gaseous $TiCl_4$ reacts with magnesium vapour, reactions that result in the formation of $TiCl_2$ and $TiCl_3$ are the most probable.

- Formation of magnesium sub-chloride (MgCl) is also a possibility at high temperatures. MgCl is unstable under normal conditions but is identified in the gaseous phase at high temperatures.

- Among the secondary reactions of $TiCl_4$ with titanium, the reaction forming $TiCl_2$ is the most probable.

S.No.	Reaction	$\Delta G^o{}_{1100K}$	$\Delta H^o{}_{1100K}$
1	½ $TiCl_4$(g) + Mg (l) = ½ Ti(s) + $MgCl_2$(l)	−155.8	−214.1
2	½ $TiCl_4$(g) + ¼ Mg(l) = ½ $TiCl_3$(g) + ¼ $MgCl_2$(l)	−38.1	−45.1
3	½ $TiCl_4$(g) + ½ Mg(l) = ½ $TiCl_2$(s/l) + ½ $MgCl_2$(l)	−168.7	−90.3
4	2/3 $TiCl_3$(g) + 1/3 Mg(l) = 2/3 $TiCl_2$(s/l) + 1/3 $MgCl_2$(l)	−60.3	−174.2
5	$TiCl_2$(l) + Mg(l) = Ti(s) + $MgCl_2$(l)	−130.9	−90.7
6	½ $TiCl_4$(g) + 1/6 Ti (s) = 2/3 $TiCl_3$(g)	−8.2	20.6
7	½ $TiCl_4$(g) + ½ Ti(s) = $TiCl_2$(s/l)	−24.9	−123.4
8	2/3 $TiCl_3$(g) + 1/3 Ti(s) = $TiCl_2$(s/l)	−16.7	−144.0

Table 1.
List of possible reactions in the Mg-Ti-Cl system.

The reduction process is carried out in a steel/stainless steel crucible/reactor under an argon gas atmosphere. It is a batch process and based on the batch size, the required quantity of magnesium is taken into the reactor and heated to 780°C. $TiCl_4$ is then pumped into the reactor from the top closure of the reactor. The temperature of reduction is maintained in the range 780–830°C by removing exothermic heat through the circulation of air around the outer surface of the reactor. Titanium particulate formed in the reduction process tends to agglomerate in sizeable pieces along the reactor's inner wall and grown pieces fall and are collected at the perforated bottom plate of the reactor vessel. The reaction by-product $MgCl_2$ (which melts at 712°C) is liquid under reactor conditions and being heavier than molten magnesium moves down paving the way for liquid magnesium to ascend to the top surface. $MgCl_2$ collected at the bottom of the reactor is periodically withdrawn from the reactor for accommodating the increased volume of the titanium sponge. The titanium sponge grows as a cylindrical cake inside the reactor. It is inevitable that some amounts of magnesium metal and magnesium chloride get entrapped in the pores of the sponge during their transportation through the sponge during the process. To compensate the loss of reductant (magnesium metal) that is entrapped in the pores, usually 50–60% excess (more than the stoichiometric requirement) magnesium is used in the sponge production campaigns. After completion of the reduction process, the sponge cake comprises un-reacted magnesium metal and magnesium chloride, which are entrapped in the pores of the titanium sponge.

Vacuum distillation of the reduced mass is carried out by taking out the reaction crucible containing the reduced mass into another reactor (**Figure 2**) and heated to about 1000°C under a dynamic vacuum of the order of 5–10 x 10^{-3} mbar, to distil out $Mg/MgCl_2$ resulting in titanium sponge freed from the entrapants. The basic principle in the vacuum distillation process is that at 1000°C, the vapour pressure of Mg and MgCl2 is higher whereas the vapour pressure of titanium is very insignificant. However, there are practical issues in the optimization of principle parameters of the distillation process such as distillation soak, temperature, monitoring of vacuum as fall in the vacuum inside the process reactor is overlapped with atmospheric leaks into the system. However, a realistic time of distillation time could be assessed through simultaneous observation of vacuum fluctuations in the process reactor/condenser vessel and electrical energy consumption pattern of heating of the reactor with reduced mass. After cooling to room temperature under vacuum/argon gas, the sponge cake is ejected out of the reactor at room temperature carefully employing appropriate equipment/tooling. **Figure 3** shows a photograph of a 3 MT sponge cake produced at the DMRL titanium research centre. Any burnt or coloured particulates (formed due to oxidation) are manually removed before the cake is further handled and processed to prepare 2–25 mm size material for taking up ingot melting.

The Kroll process has several advantages over the other two processes of sodium reduction and fused salt electrolysis in terms of scaling up product purity, and amenability to recycle magnesium metal from the by-product ($MgCl_2$) through electrolysis. Over the years there have been several technological advancements that taken place in the Kroll technology as mentioned below:

- Enlargement of batch size (from 2 to 4 MT at the beginning of the industry to 8–12 MT at present). Increased batch size is advantageous for enhanced productivity and high-quality yield.

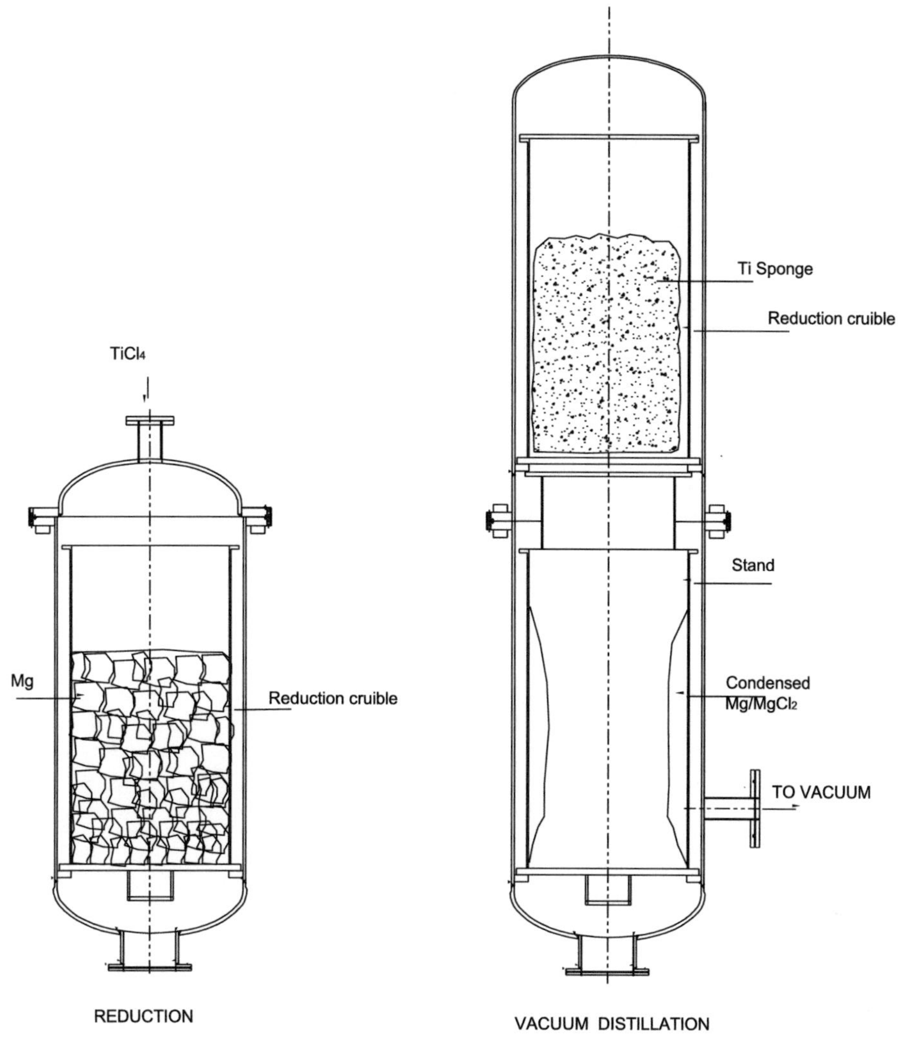

Figure 2.
Schematic of reactors for reduction and vacuum distillation processes in the conventional Kroll process.

- Combined process technology (wherein reduction and vacuum distillation processes are conducted in a single reactors assembly system (**Figure 4**). This significantly cuts down energy requirement and total cycle time of a production batch.

- Recycling of by-product $MgCl_2$ by fused salt electrolysis in multipolar cells to regenerate magnesium metal and chlorine for captive consumption in an integrated titanium sponge plant.

- Advanced process instrumentation and AI techniques for process control, sponge sorting, etc.

- Evolution of different types of reactor materials, innovative sponge quality evaluation techniques and so on.

Figure 3.
A 3 MT titanium sponge cake produced at DMRL titanium experimental facility by combined process technology.

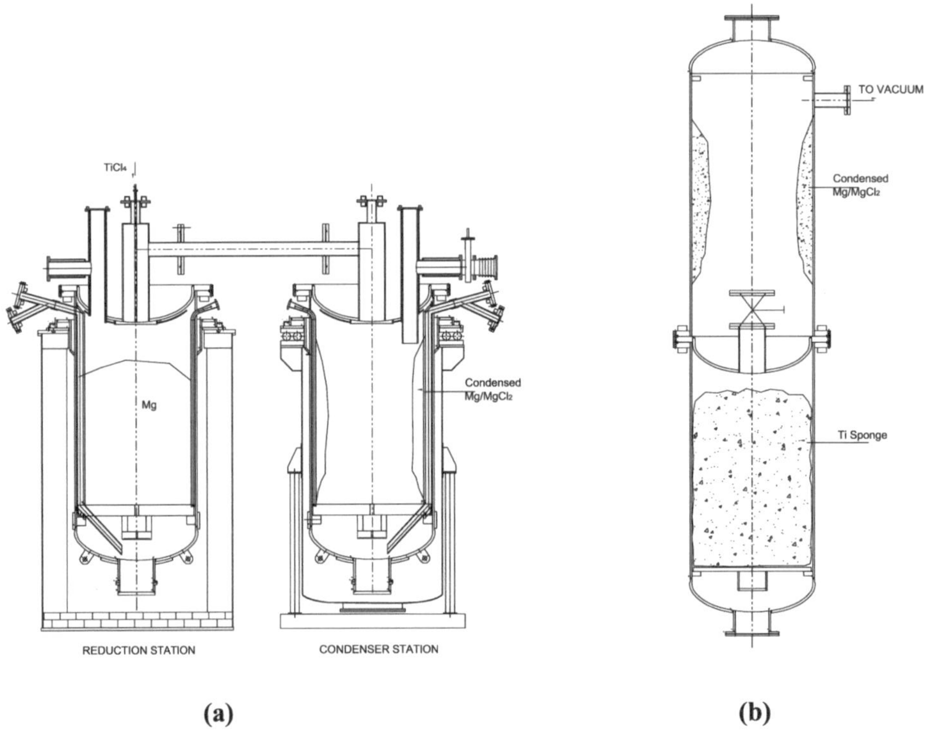

Figure 4.
Schematic arrangement of reactors in the combined process technology: (a) side by side and (b) one over the other.

6. Fused salt electrolysis of $TiCl_4$

Electrolytic processes of metal extraction have many advantages such as simplicity, scope for producing high-purity metal and amenable to semi-continuous/ continuous process operations and so on. Efforts on titanium metal extraction by fused salt electrolysis concept are very old and several researchers put in efforts to develop the same. Initially, electrolysis of TiO_2 in molten electrolytes of alkali borates and phosphates was pursued by International Research Inc., USA. However, there were many technical issues such as higher operating temperatures, corrosive electrolytes, product contamination by oxygen and difficulty in purifying the product. Subsequently, fused salt electrolysis of $TiCl_4$ in LiCl-KCl bath was considered to be more viable and extensively studied simultaneously by Imperials Chemicals Inc., UK, Dow Howmet and Reactive Metals Inc., USA, and Ginatta Marco, Italy. Consistent and sustained efforts put in by GTT, Italy, on Pilot plant studies of the electrolysis technology led to further development of the technology for industrial implementation [6, 7]. It is learnt that the multi-valency of titanium, covalent bonding in $TiCl_4$ (characterized by solubility limitation) and control of product purity have been found to be major challenges in working out a commercial model of the cell.

Among the three processes as discussed above, the Kroll process withstood the test of time and has been the predominant method of titanium sponge production the world over. Though the electrolytic process is thought to be a potential alternative to the Kroll process, to date, it remained at pilot scale operations only.

7. World production of titanium sponge

Major titanium sponge-producing countries have been the USA, Japan, Russia, Kazakhstan, Ukraine, UK, and China. In the UK, production was closed down in the early '90s. In India, a small capacity (500 MT/year) titanium sponge plant was established with the technology developed at DMRL and has been producing titanium sponges since 2012 for meeting domestic needs. Recently, Saudi Arabia started a 15,000 MT/year capacity titanium sponge plant.

The present world production capacity for titanium sponge is placed at 3,50,000 MT per annum and China is contributing about 50% of the total [8]. It is to be noted that the Kroll process is used in the entire world production.

8. Control of impurity elements in the Kroll process

In view of the wide application of the Kroll process for titanium sponge production, this chapter envisages to bring in a detailed discussion on the quality control aspects pertaining to this titanium extraction method. As already mentioned, titanium is highly reactive and the thermodynamics of the titanium-based systems do not permit any purification method that can be implemented for purifying sponge. Only control of impurities is a solution for preparing high-purity sponge. The important impurity elements in titanium sponge that adversely affect its properties are O, N, C, Fe, H, Ni, Cr, Mg and chlorides. Among this H impurity, Mg and chlorides are generally not of serious concern as they are driven out during the ingot melting.

However, Mg and chloride contents are to be kept at a minimum to improve the shelf life of the sponge (before it is taken up for ingot melting).

The following are the major factors that influence the quality/purity of titanium sponge produced by the Kroll process:

- Purity of raw materials and other consumables ($TiCl_4$, Mg, argon gas, etc.)
- Cleanliness and pressure tightness of the reactor assembly
- Process operating conditions (Reduction temperature, $TiCl_4$ feed rate, exposure of reduced mass, order of vacuum and vacuum distillation soak, overall cycle time, etc.)
- Reactor material
- Care taken during sponge cake handling and size reduction processing to prevent contamination of sponge from equipment and tooling
- Sponge storage before melting

Mastery of high-purity titanium sponge production lies in all the above parameters. For example, though the raw materials are of high purity, other parameters mentioned above tend to cause impurity elements to join the sponge. In **Table 2**, various parameters that act as a source of various impurity elements are listed. In general, sponge material adjacent to the wall gets contamination by diffusion of an impurity element from reactor material and the content of that reduces towards the inner mass. The bottom portion of the sponge cake is highly contaminated due to diffusion of impurity elements from the false bottom plate of the reactor and also, all the impurity elements present in magnesium metal transfer into the sponge before the metal ascends to the top of the liquid surface inside the reactor. The top portions of the sponge are enriched with oxygen and chloride due to atmospheric leaks from the nozzles of the reactor lid and entrainment of the last distilled chloride material. Based on the experimental data involving analysis of the large number of samples collected

S.No.	Parameter	Source of impurity elements
1.	Reactor material Carbon steel Austenitic stainless steel Ferritic chromium steel	Fe, C, O Ni, Fe, Cr Cr, Fe
2.	Raw materials TiCl4 Magnesium	O, Fe, Si, Sn, etc. Fe, Al, Si, etc.
3.	Cleanliness & pressure tightness of reactor assembly	O, N, H
4.	Process operating conditions	Fe, O, N, chloride
5.	Equipment & tooling for sponge cake processing	C, Si, O, Fe
6.	Sponge storage	O, N

Table 2.
Sources of impurities in the Kroll process of titanium sponge production.

from various sites across the sponge cake, a general pattern of impurities distribution is known. Typical distribution of impurity elements in titanium sponge is schematically shown in **Figure 5**. Thus, the sponge in the central regions/core assumes the highest purity and is separately harnessed while processing/size reduction of the sponge cake to prepare homogenous quality lots of very high purity. All titanium sponge producers evolve and follow systematic procedures for grading and quality evaluation of sponge cakes and prepare uniform quality sponge lots in finished size, which is suitable for subsequent ingot melting of Ti/Ti alloys [9].

Implementation of all quality control measures in titanium sponge production results in higher purity of the product with enhanced yield. It is claimed that industry giants prepare high pure sponges of 5 N (five nines, i.e., 99.999% Ti) and supply them for critical & electronic applications [10]. Similarly, sponge with very low iron and nickel (Fe + 3 Ni $\leq$ 120 ppm) is regularly prepared for the manufacture of critical aero-engine components.

Though ASTM MD 120 provides complete specifications on titanium sponge purity, it is advisable to consider sponges of much more purity for aerospace applications, in view of the high sensitivity of mechanical properties of titanium with respect to the impurity elements especially O, N, C and iron. Major sponge producers across the globe specify their own standards. MIDHANI is the major manufacturing unit of titanium and titanium alloys in India. In **Table 3**, MIDHANI, Japanese (SHOWA) and Russian specifications of high-purity titanium sponge are presented and a comparison is made with ASTM specification. Analysis of the best sponge produced at the DMRL research centre is also included in the table. The Brinnel Hardness Number (BHN) of a machined button melted out of a sponge sample acts as a very informative indicator of the purity of the sponge. Hence, a mention of BHN of the sponge is always found in the titanium sponge specifications.

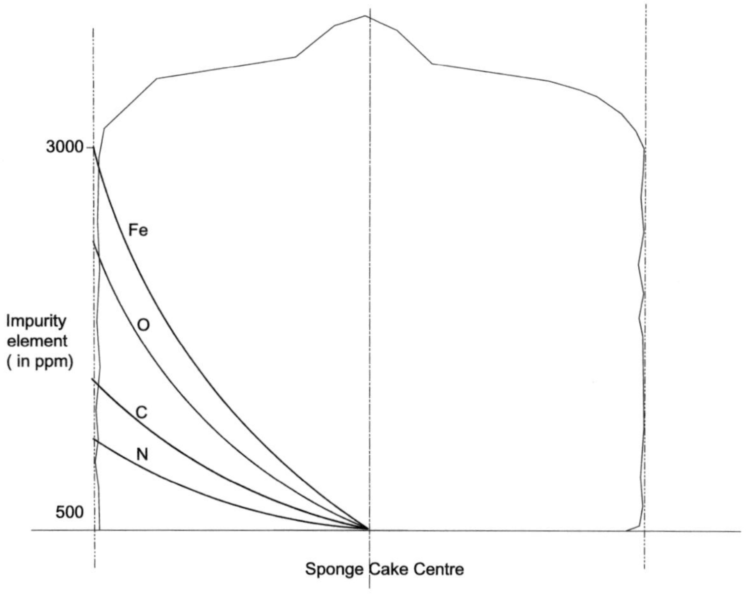

Figure 5.
Typical distribution of impurity elements in a titanium sponge cake.

Impurity element	CIS TG-90	SHOWA S-90	ASTM MD 120	DMRL lot L001/2 K	MIDHANI
O	0.040	0.060	0.100	0.032	0.080
N	0.020	0.010	0.015	0.003	0.015
C	0.030	0.020	0.020	0.006	0.015
Fe	0.060	0.030	0.120	0.018	0.050
Ni	NS	NS	NS	<0.005	0.050
Cr	NS	NS	NS	0.009	NS
Mg	0.080	0.045	0.080	0.004	0.080
Chloride	0.080	0.080	0.120	0.005	0.100
H	NS	0.002	0.001	0.002	NS
Ti (by difference)	99.6	99.8	99.6	99.8	99.6
BHN	80–90	90	120	82	100

Table 3.
Specifications of high-purity titanium sponge—different standards (content in wt%).

9. Trends in the development of alternate titanium extraction processes

The Kroll process technology though widely found acceptance has been used for industrial-scale production of titanium sponges. It has the demerits of being highly capital and labour-intensive, having higher energy consumption and higher cost of production. Improvements over the years in technology implementation have reached stagnation. Established titanium giants too claim an energy requirement of as high as 30–32 kWh/kg of sponge. However, the process has been sustaining as there is no alternative method of sponge production is evolved so far. Worldwide over there have been several efforts and intensive research continuing on alternate processes of titanium metal extraction. During the last 2–3 decades, very encouraging developments have been taking place with a few new extractions processes taken forward to even pilot-scale production. Various process technologies that are being tried out for low-cost production of titanium sponge are summarized by Kraft [11].

It is well known that titanium metal usage is restricted mainly because of higher costs. The new alternate titanium metal extraction process mainly aims at bringing down the energy requirement and cost of production. The approach for the same mainly consists of simplifying the process flow sheet by cutting down the number of unit operations, direct production of end components by combining the extraction and forming processes, etc. Thus, the processes lead to the direct production of Ti/Ti alloy powders followed by near net shape forming/3D printing (**Figure 6**) assuming greater importance and are on the way of emerging as viable alternate titanium production methods.

The following innovative processes are mentioned worthy in this context and are on the verge of adaptation for industrial-scale metal production.

9.1 The FFC (Fray Farthing & Chen) process

Titanium metal preparation directly from TiO_2 by a novel electrochemical reduction process was successfully developed and patented by Fray, Farthing and Chen of Cambridge University, UK in 1999 [12, 13]. The process essentially involves cathodic

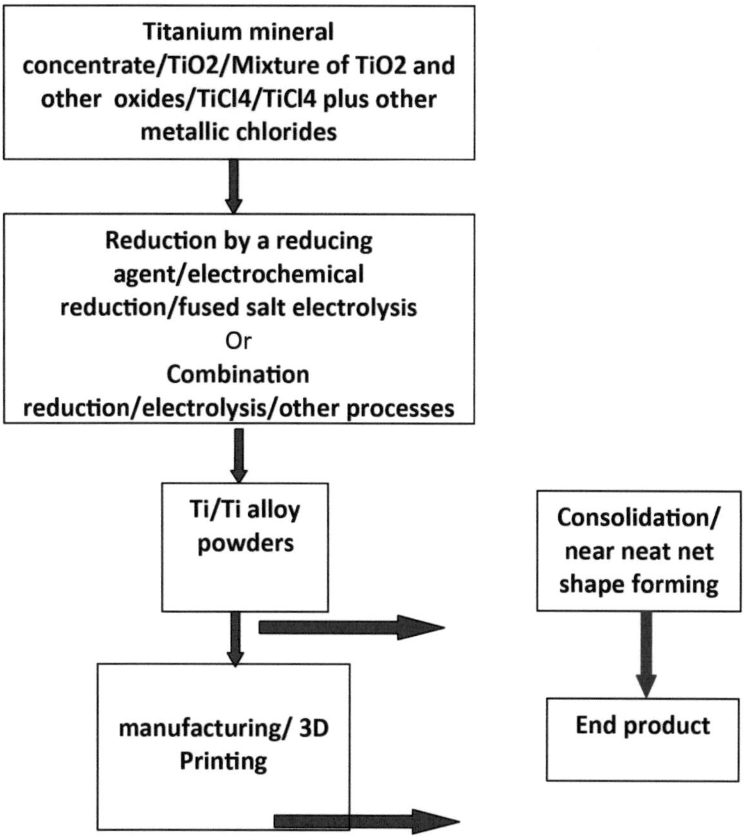

Figure 6.
Current trends in the development of alternate titanium metal extraction processes.

treatment of various types of TiO_2 pre-forms at high temperatures (950–980°C) in an argon gas atmosphere using molten calcium chloride as electrolyte and graphite as an anode. When DC voltage is applied, oxygen of the TiO_2 cathode is removed and transported through the electrolyte to the anode where CO/CO_2 is formed and let out. After the process, the sponge is subjected to water/acid leaching for removing adhered electrolyte material. Since the discovery of the process several developments/advancements has been taking place for an improved understanding of the mechanism of metallization, scaling up, enhanced product purity, applying process for alloy preparation and exploiting the principle for other possible metal oxides, etc. Basic research carried out at DMRL attempted to bring in an improved understanding of the reaction mechanism and the importance of monitoring CO_2 content in the vent gases [14, 15]. Excellent reviews by Mohan Das [16, 17] bring out details of developments that have been taking place in the FFC process over the years. Metalysis, UK, reportedly works on pilot-scale studies of the process and even exploring industrial-scale implementation [18, 19] .

10. Armstrong process

The Armstrong process is based on the sodium reduction of $TiCl_4$ (Hutner's process) with many advancements brought in to produce high-quality titanium powder particles in a continuous manner. The process technology was developed by researchers experienced in handling molten sodium at International Titanium Powder LLC, IL, USA [20]. The process technology has been demonstrated for producing pure titanium powder as well as standard titanium alloy powders. Basically, the process involves the reduction of $TiCl_4$ vapour, which is injected into a molten pool of sodium in a titanium reactor. The product of titanium powder and sodium chloride is leached with water to remove the salt. Any excess sodium metal in the product is separated by distillation.

11. OS process

Calciothermic reduction of TiO_2 with several process improvements to obviate the theoretical and practical difficulties encountered in the conventional calico-thermic reduction has been constantly pursued by Ono & Suzuki of Kyoto University, Japan. In the OS process, TiO_2 is fed to molten calcium/$CaCl_2$ bath and the reduction process is continued with the dissolution of the reduction by-product CaO in the calcium chloride. Continuous supply of the reductant, calcium metal is ensured by *in situ* electrolysis of CaO that is generated as a by-product. Pilot plant scale studies were reportedly carried out to study and establish the feasibility of continuous production of titanium metal powder [21].

12. CSIRO process

Commonwealth Scientific and Industrial Research Organization (CSIRO), Australia, pursued and experimented on gaseous phase reduction of $TiCl_4$ vapour with magnesium metal vapour (with the same reaction basis as the Kroll process) in a fluidized bed reactor [22]. The reaction by-product, $MgCl_2$ and un-reacted magnesium metal are removed by vacuum distillation, while high-purity titanium powder can be continuously withdrawn from the reactor. Based on the successful development of the process for continuous production of titanium powder, it is being taken forward to pilot-scale testing.

There are many other processes of titanium metal extraction that have been in experimentation and pursued on a lab scale. The processes include the metal hydride reduction process (MHR process) wherein TiO_2 is reduced by calcium hydride, electronically mediated reduction process (EMR) wherein TiO_2 is reduced by calcium metal without any contact of a reductant, hydrogen-assisted magnesium reduction of TiO_2 process, high temperature fused salt electrolysis of TiO_2 using metallic anode (being pursued by MIT, USA), combined magnesium thermal reduction of $TiCl_4$ electro-slag melting, etc., which promise a way ahead in the pursuit of evolving an alternated process of titanium metal extraction. A recent review on titanium production methods [23] covers many efforts that are being put in for developing a new process of titanium metal extraction. Thus, titanium extraction metallurgy provides a large scope for trying out different types of processes and techniques for producing the metal.

Acknowledgements

Author wishes to express sincere thanks and gratitude to Dr.G.Madhusudhan Reddy, Outstanding Scientist and Director, DMRL for according approval for the publication of this chapter on titanium extraction metallurgy developments. This rich experience gained by the author in the area of development & demonstration of Kroll titanium sponge production technology has been possible only because of the collective efforts of a large number of scientists, expert technical personnel and supporting staff involved in the program at DMRL. Financial support provided by DRDO, DAE and ISRO during various stages of development, demonstration and transfer of technology to the industry is gratefully acknowledged.

Author details

Chaganti R.V.S. Nagesh
Defence Metallurgical Research Laboratory, Hyderabad, India

*Address all correspondence to: nagesh.dmrl@gov.in

References

[1] McQuillan AD, McQillan MK. Titanium. London: Butterworths Scientific Publications; 1956

[2] Garmata VA. The Metallurgy of Titanium, Translation Division. Ohio: Wright Patterson Airforce Base; 1970

[3] Bomberger HB, Froes FH, Morton PH. Titanium – A Historical Perspective, in Titanium Technology-Present Status and Future Trends. USA: Titanium Development Association; 1985

[4] Subramanyam RB. In titanium –95 – Science & Technology, proceedings of 8th world conference on titanium. The Institute of Materials, UK. 1996;**2**:1551-1558

[5] Nagesh Ch RVS, Sridhar Rao C, Ballal NB, Krishna Rao P. Mechanism of titanium sponge formation in a Kroll reduction reactor. Metallurgical and Materials Transactions B. 2004;**35**:65

[6] Ginatta MV. Extractive metallurgy of primary titanium. Light Metal Age. 2004;**62**:48

[7] Ginatta MV. Economics and production of primary titanium by electrolytic winning. Procc. Of TMS Annual meeting. 2001

[8] Titanium, US Geological Survey, Mineral Commodity Summaries. 2022

[9] Nagesh CRVS, Brahmendra Kumar GVS, Saha B, Gokhale AA. Titanium sponge production and processing for aerospace applications. Aerospace Materials and Material Technologies. 2017

[10] Nakamura K, Iida T, Nakamura N, Araike T. Materials transactions. The Japanese Institute of Metals and Materials. 2017;**58**(3):319-321

[11] Kraft EH. Summary of Emerging Titanium Cost Reduction Technologies. Vancouver, WA: Report by EHK technologies; 2004

[12] Fray DJ, Farthing TW, Chen GZ. Removal of oxygen from metal oxides and solid solutions by electrolysis in a fused salt, International Patent NoWO 9964638. 1999

[13] Chen GZ, Fray DJ, Farthing TW. Direct electrochemical reduction of titanium dioxide to titanium in molten calcium chloride. Nature. 2000;**407**:361

[14] Rajulu G et al. On chemical synthesis of electrochemical reduction of titanium dioxide (TiO_2) to titanium metal. Transactions of the Indian Institute of Metals. 2016;**69**:999

[15] Rajulu G et al. Carbon dioxide (CO_2) released in the electrochemical reduction of titanium dioxide (TiO_2) to titanium metal. Materials Transactions. 2017; **58**(6):914

[16] Mohan Das KS. Direct electrochemical conversion of metal oxides to metals by molten salt electrolysis - A review. Mineral Processing and Extractive Metallurgy. 2013;**122**:95

[17] Mohan Das K et al, Molten salt based direct solid state electrochemical de-oxidation of metal oxides to metal: Our experiences at IGCAR, SMC Bulletin. vol. 10. Mumbai, India: Society for Materials Chemistry, BARC; 2019. p. 2

[18] Rao K et al. Development of new generation FFC pilot plant for production of low cost titanium & titanium alloys. In: Proceedings of 12th World Conference on Titanium, June 19-24, 2011, China. USA:

The Minerals & Metals Society (TMS); 2011

[19] Barrett R. Making the most of metal Powders. Metal Bulletin Magazine. August 2017. p. 21

[20] Grant Crowley ITPLLC. How to extract low cost titanium. Advanced Materials & Processes. 2003;**161**:25-27

[21] Suzuki RO, Teranuma KOH, Ono K. Calciothermic reduction of titanium dioxide to titanium metal and in situ electrolysis in molten CaCl2. Materials Transactions. The Japanese Institute of Metals and Materials. 2003;**34B**:287-295

[22] Doblin C, Chryss A, Monch A. Titanium Powder from the TiRO Process, in Key Engineering Materials. Stafa-Zurich, Switzerland: Trans Tech Publications; 2012

[23] Khalloufi MEI, Drevelle O, Soucy G. Titanium an over view of resources and production methods. Minerals, MDPI. 2021;**11**:1425

Section 3

Processing

Chapter 4

Low-Cost Preparation Technologies for Titanium Alloys: A Review

Qisheng Feng and Chonghe Li

Abstract

The titanium industry has been developing for nearly 70 years since the birth of Ti-6Al-4 V alloy. Due to its high specific strength, high and low-temperature resistance, corrosion resistance and good biocompatibility, titanium alloy is used in aerospace, marine engineering, and biomedical fields. However, the high production cost of titanium alloys currently limits their widespread use like steel and aluminum alloys. Therefore, the low-cost preparation technology for titanium alloys becomes hot research in recent years. This chapter provides a comprehensive overview of low-cost preparation technologies for titanium alloys from four aspects: raw materials, melting, hot working and machining, and advanced technologies. This review would be of interest to scholars in related fields.

Keywords: low-cost titanium alloys, alloying elements, low-cost melting, hot working, wire and arc additive manufacturing (WAAM)

1. Introduction

Although titanium (Ti) is the tenth amplest element on earth and the fourth most ample structural metal after aluminum (Al), iron (Fe), and magnesium (Mg), it took about 160 years from its discovery to the industrial application of Ti-6Al-4 V alloy in 1954 [1]. This is because, despite its large reserves, titanium is also defined as a rare metal due to its strong affinity for elements such as O, C, and N, which makes it difficult to extract from its ores [2]. In the past 70 years since the birth of Ti-6Al-4 V, titanium alloys have developed into hundreds of types and have been widely used in various industries due to their high specific strength, high and low-temperature resistance, corrosion resistance, and biocompatibility [3]. The strength of titanium is similar to steel, but its density is about half that of steel, and compared with aluminum, titanium is five times stronger [4]. The high specific strength of titanium alloys and their high tolerance to extreme operating conditions have led to their widespread use in aerospace (aircraft, spacecraft, engine, etc.) and military applications (helicopter, unmanned aerial vehicles, guided missiles, etc.) [5–7]. About 60% of titanium alloys are used in aerospace in Europe and the United States [8]. Nowadays, people are increasingly concerned about reducing CO_2 emissions, and the light weighting of transportation is a key step in reducing CO_2 emissions [9]. Therefore, the automotive industry is showing a strong interest in the use of titanium alloys for the light weighting of structural components, which offers the possibility of lower fuel consumption

IntechOpen

and reduced CO_2 emissions, but are currently limited to racing cars and specialized vehicles [10–12]. Titanium alloy is also used in marine engineering as a corrosion-resistant and lightweight structural material (propeller, bathyscaph, etc.) [13, 14]. In addition, non-toxic, non-magnetic, and better biocompatibility make titanium alloy widely used in medical implants (joint prosthesis, skeletal repair, etc.) [15–17]. Based on the current industrial applications and foreseeable development potential, titanium and its alloys are gradually becoming the fourth generation of metal materials after copper (Cu), iron (Fe), and aluminum (Al), but its high application cost greatly limits its large-scale application like the previous three metal materials [4, 18, 19]. Therefore, for about the past 30 years, research has been driven by various industriesto make titanium alloys a common general-purpose metal by reducing their production costs [20–23].

The traditional production and manufacturing process of titanium alloy components are roughly the same as for steel, including production of titanium sponge, addition of alloying elements, melting and pouring of alloy ingots, hot working to prepare different types of titanium materials (plate, tube, bar, etc.), and machining to prepare the final titanium alloy components. According to statistics, the percentage of each factor in the overall cost composition of titanium parts is shown in **Figure 1** [24]. It can be seen that from titanium sponge to ingot accounts for about 30% of the total cost, which includes the preparation of titanium sponge, the addition of alloying elements, and the melting of ingots. The average prices of titanium metal and its alloyed elements are shown in **Table 1** [25]. The price of titanium metal is much higher than the general metal iron (Fe) and aluminum (Al), while the application of more alloying elements, such as V, Nb, Mo, will again increase the cost of raw materials. In addition, the molten state of titanium alloy has a large viscosity and poor filling mobility, which makes it easy to produce defects such as porosity during the solidification process [26]. On the other hand, the hot working and machining accounts for about 70% of the total cost of titanium alloys, which is a hardly desirable percentage. Titanium alloys tend to react with oxygen under hot working and form a hardened oxide layer on the surface, making this process require specialized equipment [27]. At the same time, due to its low thermal conductivity, it will make the heat generated during the machining process cannot be dissipated quickly, and the higher heat accumulation will accelerate the tool wear [28]. Poor machinability and electrical conductivity have also led to the classification of

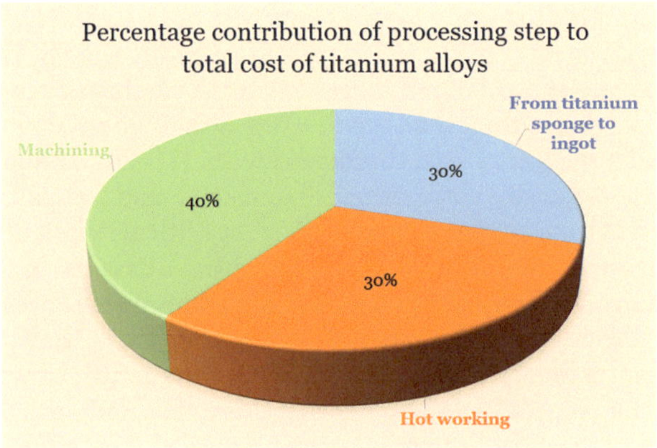

Figure 1.
Percentage contribution of processing step to total cost of titanium alloys [24].

Element	Price (USD/kg)	Role in phase stability
Ti	~11.9	
Al	~2.6	α-stabilizer
Fe	~0.6	β-stabilizer
V	~334.0	β-stabilizer
Nb	~96.5	β-stabilizer
Mo	~56.0	β-stabilizer
Ta	~534.4	β-stabilizer
Ni	~25.4	β-stabilizer
W	~16.2	β-stabilizer
Cr	~11.4	β-stabilizer
Co	~53.6	β-stabilizer
Mn	~3.5	β-stabilizer

Table 1.
The average prices of titanium metal and its alloying elements [25].

titanium alloys as difficult to machine materials [29]. The low molding rate of titanium parts and the large amount of remaining scrap being wasted are also a factor in the overall high cost of titanium alloys applications [30]. After the above analysis, we understand the various factors affecting the total cost of titanium alloys and realize that there are great challenges in the development of low-cost titanium alloys.

This chapter firstly reviewed the latest progress of low-cost titanium alloys from three aspects: raw materials, melting of alloy ingots, and processing and forming. Then, several new techniques for the low-cost preparation of titanium alloys are discussed and analyzed. Finally, the future development of low-cost titanium alloys is prospected and some suggestions are put forward.

2. Raw materials

2.1 Production of titanium

In the history of titanium production, from the birth of the Hunter process to the improved Kroll process using recyclable magnesium instead of sodium, it is still the most common method used in industrial production [31]. However, it is known from practical applications that the product obtained by the Kroll process is titanium sponge, which is six times more expensive than stainless steel production process [32]. After this problem was recognized, researchers continued to develop several new processes to reduce the cost of titanium feedstock, such as the Fray-Farthing-Chen (FFC) process [33], the Armstrong process [34], the Ono and Suzuki (OS) process [35], etc. The development of new processes for the production of titanium has been reviewed by researchers [36, 37]. This section will categorize all processes and then provide a brief overview and analysis. These new processes can be divided into two categories according to the metallurgical process: thermochemical processes and electrochemical processes. A summary of the various titanium production processes is shown in **Table 2**, including Kroll process.

To date, the vast majority of the world's titanium is produced by the Kroll process, which involves chlorination of titanium ore, purification of $TiCl_4$, magnesiothermic reduction, vacuum distillation, and electrolytic regeneration of Mg and Cl_2 [36]. We all know that the reduction of highly valence titanium compounds is a stepwise process. On the one hand, in order to minimize impurities in the final product, an excess of magnesium must be added to the magnesiothermic reduction process, which leads to a decrease in magnesium utilization. On the other hand, to ensure magnesium utilization leads to an increase of low valent titanium compounds in the final product

Processes classification	Processes name	Precursors	Advantages	Disadvantages	Refs.
Thermochemical processes	Kroll process	$TiCl_4$	Product with less oxygen content and metallic impurities	Low productivity; high cost of reductant; high energy consumption	[31]
	TiROTM process	$TiCl_4$	Continuous production	Easy introduction of oxygen	[38]
	Vapor-phase reduction process	$TiCl_4$	Continuous production	High oxygen content or high magnesium and chlorine content in Ti powder	[39]
	CSIR-Ti process	$TiCl_4$	Continuous production	Oxygen content is difficult to control	[40]
	Armstrong process	$TiCl_4$	Continuous production; excellent compressibility and denseness of Ti powder	Irregular powder morphology	[34]
	MHR process	TiO_2	Single-step reaction	High energy consumption and pollution	[41]
	PRP process	TiO_2	Highly scalable; controllable purity of Ti powder	High cost of reductant	[42]
	EMR process	TiO_2	Continuous production; high purity of Ti powder	Complicated process; difficult separation of metal and salt	[43]
	HAMR process	TiO_2	Low oxygen of Ti powder	High energy consumption	[44]
	HDH process	Ti sponge, Ti ingot, Ti mill products and Ti scrap	High purity of Ti powder; relatively inexpensive process	Irregular and angular powder morphology	[45]

Electrochemical processes	FFC process	TiO_2	Semi-continuous production; low oxygen of Ti powder	Low current efficiency; difficult separation of metal and salt	[33]
	OS process	TiO_2	Semi-continuous production; low oxygen of Ti powder	Low current efficiency; difficult separation of metal and salt	[35]
	USTB process	TiC_xO_{1-x} ($0 < x < 1$)	Semi-continuous production; low oxygen of Ti powder	Low current efficiency	[46]
	SOM process	TiO_2-containing flux	Single-step process	High process temperature	[47]

Note: CSIR: council of scientific and industrial research; MHR: metal-hydride Reduction; PRP: preform reduction process; EMR: electronically mediated reaction; HAMR: hydrogen-assisted magnesium reduction; HDH: hydride-dehydride; FFC: Fray-Farthing-Chen; OS: Ono and Suzuki; USTB: University of Science and Technology Beijing; and SOM: solid oxide membrane.

Table 2.
A summary of the various titanium production processes.

and a longer production cycle, which requires a choice between production efficiency, energy consumption and product quality. Overall, the disadvantages of Kroll process, including the expensive reducing agent Mg, long process route, low production efficiency, and high energy consumption (71.61 kWh/kg), resulted in a high price of $7.4–$10.3 per kg for titanium [4, 36]. The TiRO™ process [38], the vapor-phase reduction process [39], and the CSIR-Ti process [40] can all be seen as improvements of the Kroll process, and the Armstrom process [34] can be seen as an improvement of the Hunter process, using the same precursors and reductants, and they all attempt to develop a continuous production process with low energy consumption. Although the main precursor for titanium production is $TiCl_4$, there is also a class of processes that focus on TiO_2. Because TiO_2 can also be thermally reduced by metals, the commonly used reductants include Ca and Mg and are known as calciothermic and magnesio-thermic methods, of which the MHR process [41], PRP process [42], EMR process [43], and HAMR process [44] are typical representatives. In addition to the thermo-chemical processes mentioned above, electrochemical processes have also received a great deal of attention, and typical representatives include FFC processes [33], OS processes [35], USTB processes [46], and SOM processes [47]. **Table 2** summarizes the advantages and disadvantages of the various processes, and by looking at the corresponding references you can gain a deeper understanding of the principles and processes of each process. Although many new processes have been developed over the decades in the hope of producing titanium cost-effectively, unfortunately, none of them have yet been able to replace the Kroll process in commercial applications. In fact, if it is a mature cost-efficient new process, firstly it should reach industrial production, secondly, the quality of the titanium metal produced by this new process should be the same or even better than Kroll process, and finally, the production cost of this new process should be significantly lower than that of Kroll process. It is clear that all the new processes that have been developed so far do not meet the yield, quality, and cost targets simultaneously, and there is still a long way to go to prepare low-cost titanium alloys from this step in the production of titanium.

2.2 Addition of alloying elements

Alloying elements also have a place in the raw material cost of titanium alloys. At present, the core idea of reducing the cost of titanium alloys from alloying elements is to use Ti-6Al-4 V alloy as the target and use cheap alloying elements instead of expensive alloying elements to reduce the cost of titanium alloy, while its mechanical properties are not lower than Ti-6Al-4 V alloy. This is because Ti-6Al-4 V alloy is still the most widely used titanium alloy today [48]. A great deal of corresponding research has been conducted in the United States, Japan, and China, and many low-cost titanium alloys have been developed and some have found practical application. A summary of the low-cost titanium alloys that have been developed in various countries is shown in **Table 3**.

Depending on their influence on the β-transus temperature, the alloying elements are classified as α-stabilizers, β-stabilizers, and neutral [49]. Generally, different strengthening effects are exhibited due to different alloying elements types, but some alloying elements can be replaced by others. The α-stabilizing elements include Al, O, N, and C, with Al being the most commonly used in titanium alloy design. The β-stabilizing elements can be divided into isomorphous-stabilizing elements and eutectoid-stabilizing elements. The isomorphous-stabilizing elements include V, Mo, Nb, Ta, etc., while the eutectoid-stabilizing elements include Fe, Cr, Mn, Ni, etc. Zr and Sn are neutral elements. As can be seen in **Table 1**, eutectoid

Country	Alloy designation	Chemical composition (wt %)	Low-cost features
United States	Timetal 62S	Ti-6Al-1.7Fe-0.1Si	Replace V with Fe
	ATI-425	Ti-4A1-2.5 V-1.5Fe-0.25O	Replace V with Fe
	Timetal LCB	Ti-1.5Al-6.8Mo-4.5Fe	Replace V with Fe-Mo alloy
	RMI VM	Ti-6.4Al-1.2Fe	Replace V with Fe
	Timetal CL4	Ti-5Al-3 V-0.6Fe-0.17O	Replace V with Fe
Japan	Ti-Fe-O-N alloys	Ti-1Fe-0.35O-0.01 N Ti-1Fe-0.4O-0.045 N Ti-1Fe-0.5O-0.05 N	Replace V with Fe; replace Al with O and N
	SP-700	Ti-4.5A1-3 V-2Mo-2Fe	Replace V with Fe-Mo alloy
	TFCA	Ti-4.3Fe-7.1Cr-3.0Al	Replace V with Fe-Cr alloy
	TFC	Ti-4.3Fe-7.1Cr	Reduce Al; replace V with Fe-Cr alloy
	KS Ti-531C	Ti-4.5Al-2.5Cr-1.2Fe-0.1C	Replace V with Fe-Cr alloy
China	Ti-8LC	Ti-6Al-1Mo-1Fe	Replace V with Fe-Mo alloy
	Ti-12LC	Ti-4.5Al-1.5Fe-6.8Mo	Replace V with Fe-Mo alloy
	Ti-35,421	Ti-3Al-5Mo-4Cr-2Zr-1Fe	Replace V with Fe-Cr alloy
	Ti-5322	Ti-5Al-3 V-2Cr-2Fe	Partially replace V with Fe-Cr alloy

Table 3.
A summary of the low-cost titanium alloys in various countries.

β-stabilizing elements are cheaper than isomorphous β-stabilizing elements, so people choose to use eutectoid β-stabilizing elements instead of isomorphous β-stabilizing elements for the design of low-cost titanium alloy. For example, in the early 1960s, the Ti185 (Ti-1Al-8 V-5Fe) alloy was developed through the addition of Fe [50]. Its high tensile and shear strength makes it an attractive fastening material. Since Fe is readily available and the cheapest, it is considered to be an ideal replacement element [51]. Fe is also one of the strongest β-stabilizing elements, and studies have shown that with the addition of Fe, the grain size of Ti-Fe system alloys decreases, which can improve the comprehensive properties of alloys [52, 53]. The use of Fe completely replaces V, such as the Timetal 62S (Ti-6Al-1.7Fe-0.1Si) alloy developed in the United States [54], and the addition of Si refines the grain. The cost of this alloy is reduced by 15–20% compared with Ti-6Al-4 V alloy without loss of strength and tensile properties, and has been used in military applications. In addition, the addition of Fe can facilitate hot working by reducing flow stresses, thus saving heat and power consumption [55]. Although you have many reasons to use Fe as an additive element for low-cost titanium alloys, it has been found that when Fe is added at more than 3%, segregation of Fe tends to occur, leading to the formation of intermetallic compounds (TiFe or Ti_2Fe) and beta flecks [56]. The intermetallic compounds tend to cause the alloys to lose their plasticity and also reduce their mechanical properties, while the beta flecks are often the preferred location for fatigue failure of the alloys [57]. To avoid the above, people choose to use Fe to partially replace V and Mo as isomorphous β-stabilizing elements, such as ATI-425 alloy (Ti-4A1-2.5 V-1.5Fe-0.25O) [58], Timetal LCB alloy (Ti-1.5Al-6.8Mo-4.5Fe) [59], Ti-8LC alloy (Ti-6Al-1Mo-1Fe) [60], and Ti-12LC alloy (Ti-4.5Al-1.5Fe-6.8Mo) [60]. Timetal LCB alloy was originally developed to replace Ti-1023 (Ti-10 V-2Fe-3Al) alloy with a cheaper Fe-Mo intermediate alloy instead of V. This alloy has proven to have high strength and good formability, with mechanical properties comparable to Ti-1023 alloy while costing about 80% of Ti-6Al-4 V alloy. The Ti-8LC and Ti-12LC alloys developed by China also use a cheap Fe-Mo intermediate alloy and incorporate pure titanium scrap in the melting process. These two alloys are currently listed in the Chinese national standard, corresponding to the grades TC28 and TC29. Ti-Fe-O-N series alloy is a typical low-cost titanium alloy developed in Japan, replacing Al and V with O, N, and Fe [61]. This series of titanium alloys is generally used in civilian applications because of its low price, as well as its low plasticity and poor high temperature performance. In addition, these alloys are also considered as potential materials for biomedical applications [62]. Incidentally, low-cost biomedical titanium alloys are also being developed rapidly [16, 63]. Also as relatively inexpensive eutectoid β-stabilizing elements Cr and Mn are of course used in the development of low-cost titanium alloys, such as Ti-3Al-2Fe-8.5Cr [64], Ti-6.0Al-4.5Cr-1.5Mn [65], and Ti-4.5Al-6.9Cr-2.3Mn [66].

Because the proportion of alloying elements in titanium alloys is relatively low, compared with the main element Ti, this method is limited in terms of cost reduction. During the literature search, it is clear that although researchers in various countries have designed a large number of low-cost titanium alloys and claimed that they have good prospects for application, very few of them are actually used. The authors believe that it is time to conduct in-depth fundamental research on the process, organization, and property relationships of the already designed low-cost titanium alloys to bring them to the level of industrial application as far as possible. Take Ti-5321 (Ti-5Al-3Mo-3 V-2Zr-2Cr-1Nb-1Fe) alloy as an example, since it was designed, developers have conducted extensive research on the alloy in terms of plane strain fracture behavior, and impact toughness, fracture toughness, deformation mechanism, and mechanical properties, which can lay a good foundation for pilot scale and even industrial applications [67–73].

3. Melting of alloy ingots

3.1 Recycling of titanium scraps

The recycling and reuse of titanium scrap can reduce the production cost of titanium alloys, which means that the actual utilization rate of the material is improved. Titanium scraps are first generated from the production of titanium metal. For example, when titanium sponge is produced using the Kroll process, 10–20% of inferior titanium sponge is produced during the reduction and separation process [30]. In addition, a large amount of titanium scraps will be generated during the processing of titanium alloy parts, especially in the aircraft industry, the finished product rate of general aviation parts is only about 10% [20]. It is obvious that recycling of titanium scrap is necessary, and according to statistics, the cost of titanium alloy ingots will be reduced by 0.7% for every 1% of titanium added after the recycling process [20]. These reviews [30, 74–77] are very useful for you to get a comprehensive understanding of the current technology and status of recycling of titanium scraps, and the authors will not repeat them here. Titanium scraps that meet the reuse criteria will be mentioned in Section 3.2.

3.2 Low-cost melting technologies

The low-cost melting of titanium alloy is mainly considered from two aspects, one is to increase the utilization of titanium scraps, and the other is to improve the melting efficiency and quality, while achieving the integration of meltage and refinement. The CHM (Cold Hearth Melting) technology developed in the United States basically solves both problems [20], including EBCHM (Electron Beam Cold Hearth Melting) technology and PACHM (Plasma Arc Cold Hearth Melting) technology. The biggest difference between the above two melting technology is the difference in heat source, and one is selected here for introduction. Compared with conventional vacuum arc remelting, EBCHM has the following outstanding advantages [20, 78]: (1) High-density impurities, low-density impurities, and volatile impurities can be removed to prepare pure titanium alloys; (2) low raw material requirements, 100% titanium scraps, and titanium alloys scraps can be utilized; (3) ingots can be produced by a single melting; and (4) the preparation of multiple forms of ingots can be achieved by adjusting the crystallizer to shorten the subsequent process flow. Since EBCHM works under high vacuum conditions, alloying elements (e.g., Al) are easily evaporated if their vapor pressure is higher than that of titanium. It is particularly important to control the content of aluminum, which is a key element in most titanium alloys. Using Langmuir equation and penetration theory, the evaporation process of Al during EBCHM of Ti64 alloy was calculated and the evaporation mechanism of Al was analyzed, and the results showed that the evaporation of Al is a double-controlled process, that is, surface evaporation and melt pool diffusion [78]. To control the evaporation of Al, the surface temperature of the alloys must be controlled. The influence of melting speed, electron beam output power, and other factors on ingot composition was determined by developing a mathematical model of Al evaporation kinetics during EBCHM, which provides a usable tool for melting ingots with specified compositions using EBCHM [79]. The aerospace materials standard of the United States requires aerospace structural parts made of titanium alloys must undergo an EBCHM melting. EBCHM is able to make full use of titanium scrap and obtain homogeneous titanium ingots by primary melting, thus reducing the production cost of titanium ingots in terms of both raw materials and energy, and has become the first choice of countries.

4. Hot working and machining

As shown in **Figure 1**, from titanium metal to alloy ingots accounts for 30% of the total cost of titanium alloy parts, which means that the savings from reducing the production cost of titanium metal, using cheap alloying elements instead of expensive alloying elements and using low-cost melting technology will only add up to much less than 30%. Correspondingly, the hot working and machining processes used for the final forming of titanium alloy parts account for approximately 70% of the total cost. Obviously, it is a big challenge to reduce the cost of hot working and machining. Researchers first thought of tackling this challenge by optimizing the process parameters. The appropriate hot working parameters for different titanium alloys need to be found accurately; otherwise, poorly performing titanium alloys will only increase the cost of their applications in disguise [80]. Temperature, strain rate, and strain are the most important parameters in the hot working. Titanium alloys are sensitive to the process parameters of hot working, and changes in parameters can alter their microstructural morphology, which in turn can lead to significant changes in actual alloys properties [81, 82]. In recent years, constitutive models [83] and processing maps [84] have gained more practical applications due to their accuracy in process parameter optimization. **Table 4** summarizes some studies on constitutive models and processing maps of titanium alloys during hot working. From the different studies presented in **Table 4**, it can be concluded that the main challenge of this approach is to select the most suitable constitutive models to accurately describe the stress-strain behavior during

Alloys	Parameters			Refs.
	Temperature (°C)	Strain rate (s^{-1})	Strain (%)	
Ti-6Al-4 V	900–1050	0.1, 1 and 10	80	[85]
Ti60	970–1120	0.01–10	80	[86]
CP-Ti	400–700	0.001–1	90	[87]
IMI-834	975–1100	0.1 and 1	80	[88]
Ti2448	750–850	0.001–63	70	[89]
Ti-6Al-4 V	800–1100	0.001–10	100	[90]
BT3-1	800–1060	0.0003–1	60	[91]
As-cast ATI 425	950–1150	0.001–1	50	[92]
Ti-3Al-3.7Cr-2Fe-0.1B	800–950	0.01–10	70	[93]
46Ti-46Al-4Nb-2Cr-2Mn	850–1050	0.0001–10	80	[94]

Table 4.
A summary of some studies on constitutive models and processing maps of titanium alloys during hot working.

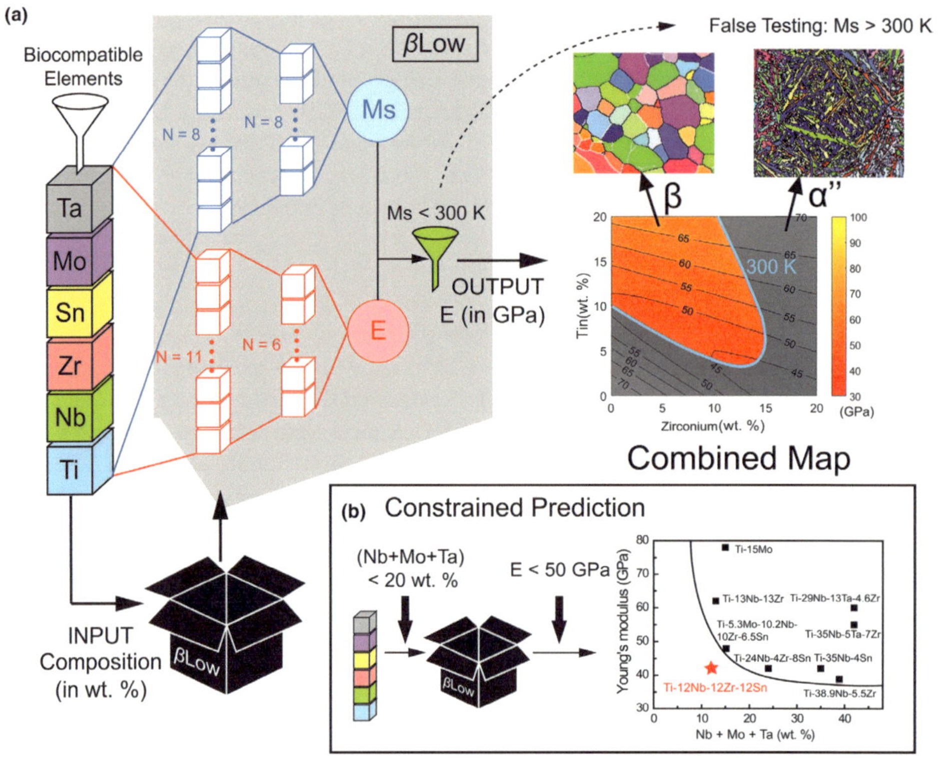

Figure 2.
Schematic diagram of artificial neural network assisted development of titanium alloys. (a) The operation process of alloy design is introduced in detail; and (b) prediction of low-cost titanium alloys by limiting high-cost elements to less than 20% [99].

the hot working. In addition, artificial neural networks based on machine learning have been applied to predict the hot deformation behavior [95], flow stress [96], microstructure evolution [97], and processing parameters [98] of titanium alloys. For example, a low-cost titanium alloy with bone-like Young's modulus was designed using neural networks, as shown in **Figure 2** [99]. And its Young's modulus and Ms. temperatures were predicted, and further experiments proved its high tensile strength and better biocompatibility. The process of titanium alloys development can be accelerated through the use of artificial neural networks, while reducing research costs.

If the cost of equipment is considered, steel and titanium co-production can be implemented [4], which can effectively use the existing mature equipment to greatly reduce the processing and manufacturing cost of titanium alloys.

5. Advanced low-cost preparation technologies

A large part of the high price of titanium alloys comes from the long and complex process from ingot to product. Advanced preparation technologies are dedicated to the preparation of titanium alloy products in a few simple steps, while achieving efficient manufacturing and improved raw material utilization. **Table 5** summarizes some studies of advanced low-cost preparation techniques of titanium alloys. The advanced technologies in the table can be divided into three categories, namely AM,

Alloys	Feedstock	Technologies	Refs.
Ti-xCu	Ti and Cu spherical powders	AM	[100]
Ti-xFe	HDH Ti powder and Fe powder	PM	[101]
Ti-6Al-4 V	Ti-6Al-4 V HDH powder	FAST-forge	[102]
Ti-xFe	Ti and Fe power	PM	[103]
Ti-xFe	HDH Ti powder and AISI 430 LHC powder	PM	[104]
Ti-35Nb	Ti and Nb power	SLM	[105]
Ti-6Al-4 V	Ti-6Al-4 V wire	WAAM	[106]

Note: AM: additive manufacturing; PM: powder metallurgy; FAST: field-assisted sintering technology; SLM: selective laser melting; and WAAM: wire and arc additive manufacturing.

Table 5.
A summary of some studies of advanced low-cost preparation techniques of titanium alloys.

PM, and FAST-forge, as SLM and WAAM are included in AM [105, 106]. The feedstock used in all technologies is metal powders, except for WAAM, which uses metal wire as feedstock. Compared with titanium sponge powder and titanium powder prepared by other methods in Section 2.1, HDH titanium powder containing Ti mill products and Ti scrap is preferred due to its low cost, low chlorine content, and good mechanical properties [45]. We note that Fe remains an important alloying element in advanced technologies, due to the fact that Fe, in addition to the advantages mentioned in Section 2.2, and helps to improve the sinterability of titanium alloys, making it more advantageous in PM [24]. Although alloying elements such as Ni, Cr and Cu are also used in powder form, the attraction of using Fe powder in advanced technologies is still the greatest, and expensive alloying elements, complex hot working, and difficult machining can all be solved at once.

AM is a process of manufacturing parts layer by layer and is a promising method for manufacturing near final (net) shape parts [100], which focuses on reducing the production cost of titanium alloy parts in terms of both improving the utilization of raw materials and preparing the final product in a short process. In the following, WAAM will be discussed and analyzed as an example. A complete WAAM process consists of three main aspects: the planning of the target part by software, the deposition of metal materials using the WAAM system, and post-treatment [107]. The cost data for the production of titanium alloy (Ti-6Al-4 V) parts using WAAM are shown in **Table 6** [108]. The capital cost is a one-time cost of approximately \$130,000 to make up the basic hardware of WAAM. Also, welding wire (Ti-6Al-4 V) is an inexpensive feedstock, with prices ranging from \$120 per kg to \$300 per kg (depending upon wire diameter). The advantages of WAAM are high deposition rates, low equipment costs, and high material utilization. Compared with conventional subtractive manufacturing, the WAAM process can reduce manufacturing time by 40–60% and post-machining time by 15–20%, and save approximately 78% of raw materials [109]. In aerospace, the "Buy-To-Fly" (BTF) ratio is used as a cost lever to refer to the ratio of materials purchased to materials in the final product, and it is clear that a lower BTF ratio is advantageous [110]. The researchers compared the BTF ratio using WAAM with conventional titanium machining and can conclude that the BTF ratio of WAAM is significantly lower compared with conventional titanium machining methods [111–113]. The applications of titanium alloy parts prepared by WAAM are shown in **Figure 3** [107]. As an example,

Costs	Items	Amount of money (USD)
Capital cost	Six-axis robot	~60,000
	Power source and torch	~36,000
	Clamping tool	~12,000
	Enclosure	~24,000
Software cost	WAAMsoft	/
Material cost	Welding wire (Ti-6Al-4 V)	~120–300/kg

Table 6.
A summary of cost data for the production of titanium alloy (Ti-6Al-4 V) parts using WAAM [108].

WAAM3D Company produced a full-scale sample of a Ti-6Al-4 V pressure container for space exploration that is approximately 1 meter high and 8.5 kg in weight. By using WAAM, 65% time was saved and 80% less raw material was used to meet performance requirements [107]. By far, WAAM products are the most commonly used in the aerospace and nuclear industries. The low processing accuracy and large surface roughness limit its application in the electronics and biomedical industries compared with other metal powder-based AM, such as SLM [114].

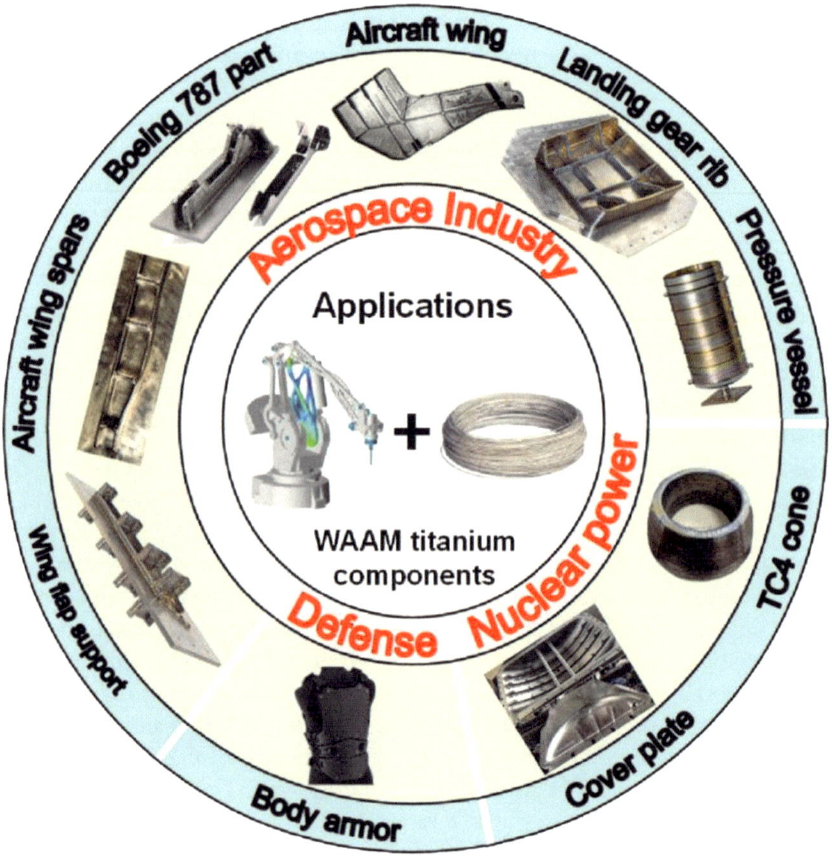

Figure 3.
The applications of titanium alloy parts prepared by WAAM [107].

6. Emerging technologies and climate

The greenhouse effect caused by CO_2 emissions is becoming increasingly evident. According to estimates, to achieve a global temperature rise limited to less than 2°C by 2100, CO_2 emissions must be reduced to less than 500 million tons per year before 2050 [115]. In recent years, several countries have developed a series of policies to address CO_2 emissions, such as carbon taxes and carbon trading schemes, all aimed at addressing the problem of climate degradation due to excessive CO_2 emissions [116–118]. It is clear that these policies will lead to changes in the industry, making the future of industrial production cleaner and more efficient. For the main process used to produce titanium, the Kroll process, researchers compared the environmental impacts of advanced technology equipment before and after its application, such as multi-pole electrolytic cells and inverted U-shaped reduction distillation furnaces [119]. The results show that the application of advanced technologies significantly improved the Kroll process' ability to save energy and reduce emissions, with 17% reduction in total electric power consumption, 90% reduction in particulate matter and CO gas emissions, and 21% reduction in overall environmental impact. In China, 74.4% of electricity is generated by coal, so a reduction in energy consumption in industrial production can also be seen as reducing carbon emissions [120]. Electrical discharge machining (EDM), which has the advantages of non-contact machining, non-macro-cutting force and high-machining hardness materials, has been maturely applied to the machining of titanium alloys [121]. According to the World Research Institute/World Business Council for Sustainable Development Greenhouse Gas Protocol, the total carbon emissions of the electricity used by an EDM machine are 23.57 tons of CO_2 equivalent. In order to save energy and improve EDM productivity, researchers developed a magnetic field-assisted EDM process (MF-EDM), and experimental results showed 61.43% reduction in energy consumption when using MF-EDM, equivalent to 9.09 tons reduction in CO_2 emissions [120]. Environmental friendliness is one of the major advantages of electrochemical methods. Unlike the preparation of titanium using electrochemical methods in Section 2.1, researchers used electrochemical methods to efficiently prepare Ti-Al alloy powders [122] and Ti-Al-V alloy powders [123] in one step, and at the same time, these titanium alloy powders can be directly used as raw materials for near-net forming of titanium alloy parts, such as powder metallurgy (PM) and additive manufacturing (AM). Through modeling and estimation, if the U.S. aircraft industry were to shift from conventional manufacturing (CM) to additive manufacturing (AM), cumulative energy savings of 1.2–2.8 billion GJ could be achieved in 2050, with associated cumulative greenhouse gases (GHG) reductions estimated at 92.1–215.0 million metric tons [112]. For any step of the full production process of titanium alloy parts, emerging technologies with environmental friendliness, low energy consumption, and low cost are the subjects of our research, and it is clear that we still have a long way to go.

7. Conclusions

In order to expand the application range of titanium and titanium alloys, the low-cost titanium alloys are one of the hot spots of research in the field of titanium alloys at present. This chapter provides a comprehensive overview of low-cost preparation technologies for titanium alloys from four aspects: raw materials, melting, hot working and machining, and advanced technologies. Raw materials include the production of titanium metal and the addition of alloying elements. Although there are many improved or alternative processes for the production of titanium, the Kroll process is still the most used in industry. Therefore, using cheap alloying elements instead of expensive ones is currently an effective way to reduce raw material costs. On the melting side, the recycling of titanium scraps and the application of CHM melting technology are together influencing the cost of alloy ingots. The use of the constitutive model, processing maps, and artificial neural networks can make the process parameters of hot working more accurate, and the co-production of steel and titanium can reduce the production cost of titanium alloys from equipment. Advanced preparation techniques for titanium alloys, such as AM and PM, are highly attractive because they generally have simple steps and can address multiple factors that lead to high costs in traditional titanium alloy preparation processes at once. Only technological progress in many aspects will enable titanium alloys to be used more widely, like steel and aluminum alloys. In addition, some emerging technologies for the preparation of titanium alloys and their contribution to the reduction of CO_2 emissions and energy consumption are presented.

Acknowledgements

This work was supported by the National Natural Science Foundation of China (Contract: U1860203), the Shanghai Rising-Star Program (19QA1403600), and the Program for Professor of Special Appointment (Eastern Scholar) at Shanghai Institutions of Higher Learning (TP2019041).

Conflict of interest

The authors declare no conflict of interest.

Author details

Qisheng Feng[1,2] and Chonghe Li[1,2*]

1 State Key Laboratory of Advanced Special Steel and Shanghai Key Laboratory of Advanced Ferrometallurgy and School of Materials Science and Engineering, Shanghai University, Shanghai, China

2 Shanghai Special Casting Engineering Technology Research Center, Shanghai, China

*Address all correspondence to: chli@staff.shu.edu.cn

References

[1] Leyens C, Peters M. Titanium and Titanium Alloys: Fundamentals and Application. Weinheim: Wiley-VCH; 2003. DOI: 10.1002/3527602119

[2] Nie X, Dong L, Bai C, Chen D, Qiu G. Preparation of Ti by direct electrochemical reduction of solid TiO2 and its reaction mechanism. Transactions of Nonferrous Metals Society of China. 2006;**16**:723-727. DOI: 10.1016/S1003-6326(06)60288-4

[3] Fashu S, Lototskyy M, Davids M, Pickering L, Linkov V, Sun T, et al. A review on crucibles for induction melting of titanium alloys. Materials & Design. 2020;**186**:108295. DOI: 10.1016/j.matdes.2019.108295

[4] Qiu G, Guo Y. Current situation and development trend of titanium metal industry in China. International Journal of Minerals, Metallurgy and Materials. 2022;**4**:599-610. DOI: 10.1007/s12613-022-2455-y

[5] Banerjee D, Williams J. Perspectives on titanium science and technology. Acta Materialia. 2013;**61**:844-879. DOI: 10.1016/j.actamat.2012.10.043

[6] Boyer R. Titanium for aerospace: Rationale and applications. Advanced Performance Materials. 1995;**2**:349-368. DOI: 10.1007/BF00705316

[7] Boyer B. An overview on the use of titanium in the aerospace industry. Materials Science and Engineering: A. 1996;**213**:103-114. DOI: 10.1016/0921-5093(96)10233-1

[8] Gomez-Gallegos A, Mandal P, Gonzalez D, Zuelli N, Blackwell P. Studies on titanium alloys for aerospace application. Defect and Diffusion Forum. 2018;**385**:419-423. DOI: 10.4028/www.scientific.net/DDF.385.419

[9] Raupach M, Marland G, Ciais P, Quere C, Canadell J, Klepper G, et al. Global and regional drivers of accelerating CO_2 emissions. PNAS. 2007;**104**:10288-10293. DOI: 10.1073/pnas.0700609104

[10] Hartman A, Gerdemann S, Hansen J. Producing lower-cost titanium for automotive applications. Journal of Metals. 1998;**50**:16-19. DOI: 10.1007/S11837-998-0408-1

[11] Takahashi K, Mori K, Takebe H. Application of titanium and its alloys for automobile parts. MATEC Web of Conferences. 2020;**321**:02003. DOI:10.1051/MATECCONF/202032102003

[12] Froes F, Friedrich H, Kiese J, Bergoint D. Titanium in the family automobile: The cost challenge. Journal of Metals. 2004;**56**:40-44. DOI: 10.1007/S11837-004-0144-0

[13] Gorynin I. Titanium alloys for marine application. Materials Science and Engineering: A. 1999;**263**:112-116. DOI: 10.1016/S0921-5093(98)01180-0

[14] Rohith K, Shreyas S, Appaiah K, Sheshank R, Ganesha B, Vinod B. Recent material advancement for marine application. Materials Today: Proceedings. 2019;**18**:4854-4859. DOI: 10.1016/j.matpr.2019.07.476

[15] Zhang L, Chen L. A review on biomedical titanium alloys: recent progress and prospect. Advanced Engineering Materials. 2019;**21**:1801215. DOI: 10.1002/adem.201801215

[16] Sidhu S, Singh H, Gepreel M. A review on alloy design, biological

response, and strengthening of β-titanium alloys as biomaterials. Materials Science and Engineering: C. 2021;**121**:111661. DOI: 10.1016/j.msec.2020.111661

[17] Gepreel M, Niinomi M. Biocompatibility of Ti-alloys for long-term implantation. Journal of the mechanical behavior of biomedical materials. 2013;**20**:407-415. DOI: 10.1016/j.jmbbm.2012.11.014

[18] Devaraj A, Joshi V, Srivastava A, Manandhar S, Moxson V, Duz V, et al. A low-cost hierarchical nanostructured beta-titanium alloy with high strength. Nature Communications. 2016;**7**:11176. DOI: 10.1038/ncomms11176

[19] Froes F, Gungor M, Imam M. Cost-affordable titanium: The component fabrication perspective. Journal of Metals. 2007;**59**:28-31. DOI: 10.1007/S11837-007-0074-8

[20] Ogawa M. Research and development of low cost titanium alloys. Journal of Japan Institute of Light Metals. 2005;**55**:549-552. DOI: 10.2464/jilm.55.549

[21] Barnes J, Peter W, Blue C. Evaluation of low cost titanium alloy products. Materials Science Forum. 2009;**618-619**:165-168. DOI: 10.4028/www.scientific.net/MSF.618-619.165

[22] Niinomi M, Nakai M, Itsumi Y, Murakami S, Oyama H, Abe W, et al. Microstructural control and mechanical properties of low-cost titanium alloys for next generation aircrafts. In: Proceedings of the 8th Pacific Rim International Congress on Advanced Materials and Processing (PRICM-8). Hawaii, USA: Wiley-VCH; 2013. pp. 1131-1138

[23] Niinomi M, Nakai M, Hieda J, Cho K, Akahori T, Hattori T, et al. Research and development of low-cost titanium alloys for biomedical applications. Key Engineering Materials. 2013;**551**:133-139. DOI: 10.4028/www.scientific.net/KEM.551.133

[24] Bodunrin M, Chown L, Omotoyinbo J. Development of low-cost titanium alloys: A chronicle of challenges and opportunities. Materialstoday: Proceedings. 2021;**38**:564-569. DOI: 10.1016/j.matpr.2020.02.978

[25] Available from: https://www.ccmn.cn/ [Accessed: June 16, 2022]

[26] Sung S, Kim Y. Melting and casting of titanium alloys. Materials Science Forum. 2007;**539-543**:3601-3606. DOI: 10.4028/www.scientific.net/MSF.539-543.3601

[27] Zhang R, Wan Y, Ai X, Liu Z. Bioactive film generation during titanium alloy cutting in oxygen-enriched atmosphere. Materials Science Forum. 2014;**800-801**:37-41. DOI: 10.4028/www.scientific.net/MSF.800-801.37

[28] Zanger F, Schulze V. Investigations on mechanisms of tool wear in machining of Ti-6Al-4V using FEM simulation. Procedia CIRP. 2013;**8**:158-163. DOI: 10.1016/j.procir.2013.06.082

[29] Thomas M, Turner S, Jackson M. Microstructural damage during high-speed milling of titanium alloys. Scripta Materialia. 2010;**62**:250-253. DOI: 10.1016/j.scriptamat.2009.11.009

[30] Takeda O, Okabe T. Current status of titanium recycling and related technologies. Journal of Metals. 2018;**71**:1981-1990. DOI: 10.1007/S11837-018-3278-1

[31] Zhang Y, Fang Z, Sun P, Zheng S, Xia Y, Free M. A perspective on thermochemical and electrochemical

processes for titanium metal production. Journal of Metals. 2017;**69**:1861-1868. DOI: 10.1007/S11837-017-2481-9

[32] Seetharaman S, Mclean A. Treatise on process metallurgy. In: Industrial Processes. Vol. 3. Amsterdam: Elsevier; 2014. DOI: 10.1016/c2010-0-67121-5

[33] Chen G, Fray D, Farthing T. Direct electrochemical reduction of titanium dioxide to titanium in molten calcium chloride. Nature. 2000;**407**:361-364. DOI: 10.1038/35030069

[34] Araci K, Mangabhai D, Akhtar K. Production of titanium by the Armstrong Process®. Titanium Powder Metallurgy. 2015;**9**:149-162. DOI: 10.1016/B978-0-12-800054-0.00009-5

[35] Suzuki R, Ono K, Teranuma K. Calciothermic reduction of titanium oxide and in-situ electrolysis in molten $CaCl_2$. Metallurgical and Materials Transactions B. 2003;**34**:287-295. DOI: 10.1007/S11663-003-0074-1

[36] Reddy R, Shinde P, Liu A. Review—The emerging technologies for producing low-cost titanium. Journal of The Electrochemical Society. 2021;**168**:042502. DOI: 10.1149/1945-7111/ABE50D

[37] Fang Z, Paramore J, Sun P, Chandran K, Zhang Y, Xia Y, et al. Powder metallurgy of titanium—Past, present, and future. International Materials Reviews. 2018;**63**:407-459. DOI: 10.1080/09506608.2017.1366003

[38] Doblin C, Chryss A, Monch A. Titanium powder from the TiRO™ process. Key Engineering Materials. 2012;**520**:95-100. DOI: 10.4028/www.scientific.net/KEM.520.95

[39] Hansen D, Gerdemann S. Producing titanium powder by continuous vapor-phase reduction. Journal of Metals. 1998;**50**:56-58. DOI: 10.1007/S11837-998-0289-3

[40] Vuuren D, Oosthuizen S, Heydenrych M. Titanium production via metallothermic reduction of $TiCl_4$ in molten salt: Problems and products. Journal of The South African Institute of Mining and Metallurgy. 2011;**111**:141-148. DOI: 10.1155/2011/464916

[41] Froes F. The production of low-cost titanium powders. Journal of Metals. 1998;**50**:41-43. DOI: 10.1007/S11837-998-0413-4

[42] Okabe T, Oda T, Mitsuda Y. Titanium powder production by preform reduction process (PRP). Journal of Alloys and Compounds. 2004;**364**:156-163. DOI: 10.1016/S0925-8388(03)00610-8

[43] Park I, Abiko T, Okabe T. Production of titanium powder directly from TiO_2 in $CaCl_2$ through an electronically mediated reaction (EMR). Journal of Physics and Chemistry of Solids. 2005;**66**:410-413. DOI: 10.1016/j.jpcs.2004.06.052

[44] Zhang Y, Fang Z, Xia Y, Sun P, Devener B, Free M, et al. Hydrogen assisted magnesiothermic reduction of TiO_2. Chemical Engineering Journal. 2017;**308**:299-310. DOI: 10.1016/j.cej.2016.09.066

[45] McCracken C, Barbis D, Deeter R. Key characteristics of hydride-dehydride titanium powder. Powder Metallurgy. 2011;**54**:180-183. DOI: 10.1179/174329011X13045076771849

[46] Jiao S, Zhu H. Novel metallurgical process for titanium production. Journal of Materials Research. 2006;**21**:2172-2175. DOI: 10.1557/JMR.2006.0268

[47] Suput M, Delucas R, Pati S, Ye G, Pal U, Powell A. Solid oxide membrane

technology for environmentally sound production of titanium. Mineral Processing and Extractive Metallurgy. 2008;**117**:118-122. DOI: 10.1179/174328508X290911

[48] Lu K. The future of metals. Science. 2010;**328**:319-320. DOI: 10.1126/science.1185866

[49] Lu J, Ge P, Zhao Y. Recent development of effect mechanism of alloying elements in titanium alloy design. Rare Metal Materials and Engineering. 2014;**43**:775-779. DOI: 10.1016/S1875-5372(14)60082-5

[50] Joshi V, Lavender C, Moxon V, Duz V, Nyberg E, Weil K. Development of Ti-6Al-4V and Ti-1Al-8V-5Fe alloys using low-cost TiH_2 powder feedstock. Journal of Materials Engineering and Performance. 2013;**22**:995-1003. DOI: 10.1007/s11665-012-0386-x

[51] Esteban P, Ruiz-Navas E, Bolzoni L, Gordo E. Low-cost titanium alloys? Iron may hold the answers. Metal Powder Report. 2008;**63**:24-27. DOI: 10.1016/S0026-0657(09)70040-2

[52] Hsu H, Hsu S, Wu S, Lee C, Ho W. Structure and mechanical properties of as-cast Ti-5Nb-xFe alloys. Materials Characterization. 2010;**61**:851-858. DOI: 10.1016/j.matchar.2010.05.003

[53] Lin D, Lin J, Ju C. Structure and properties of Ti-7.5Mo-xFe alloys. Biomaterials. 2002;**23**:1723-1730. DOI: 10.1016/S0142-9612(01)00233-2

[54] Montgomery J, Wells M, Roopchand B, Ogilvy J. Low-cost titanium armors for combat vehicles. Journal of Metals. 1997;**49**:45-47. DOI: 10.1007/BF02914684

[55] Liang Z, Miao J, Brown T, Sachdev A, Williams J, Luo A. A low-cost and high-strength Ti-Al-Fe-based cast titanium alloy for structural applications. Scripta Materialia. 2018;**157**:124-128. DOI: 10.1016/j.scriptamat.2018.08.005

[56] Zhu C, Peng G, Lin Y, Zhang X, Liu C, Zhou K. Effects of Mo and Cr contents on microstructures and mechanical properties of near β-Ti alloy. Materials Science and Engineering: A. 2021;**825**:141882. DOI: 10.1016/j.msea.2021.141882

[57] Ng C, Bermingham M, Yuan L, Dargusch M. Towards β-fleck defect free additively manufactured titanium alloys by promoting the columnar to equiaxed transition and grain refinement. Acta Materialia. 2022;**224**:117511. DOI: 10.1016/j.actamat.2021.117511

[58] Boyer R, Williams J. Developments in research and applications in the titanium industry in the USA. In: Proceedings of the 12th World Conference on Titanium (Ti-2011). Beijing, China: Science Press Beijing; 2012. pp. 10-19

[59] Kosaka Y, Fox S, Faller K, Reichman S. Properties and processing of TIMETAL LCB. Journal of Materials Engineering and Performance. 2005;**14**:792-798. DOI: 10.1361/105994905X75637

[60] Abdalla A, Amrin A, Muhamad S, Hanim M. Development of iron (Fe) additions approach for cost reduction in Ti-alloy—A review. International Journal of Materials and Product Technology. 2017;**55**:56-73. DOI: 10.1504/IJMPT.2017.084975

[61] Ikeda M, Ueda M. Development and research of low-cost titanium alloys, especially case of Japan. Materials Science Forum. 2016;**879**:119-124. DOI: 10.4028/www.scientific.net/MSF.879.119

[62] Koike M, Ohkubo C, Sato H, Fujii H, Okabe T. Evaluation of cast Ti-Fe-O-N

alloys for dental applications. Materials Science and Engineering: C. 2005;**25**:349-356. DOI: 10.1016/j.msec.2005.04.002

[63] Abd-elrhman Y, Gepreel M, Abdel-Moniem A, Kobayashi S. Compatibility assessment of new V-free low-cost Ti-4.7Mo-4.5Fe alloy for some biomedical applications. Materials & Design. 2016;**97**:445-453. DOI: 10.1016/j.matdes.2016.02.110

[64] Wang J, Qin Z, Xiong F, Wang S, Lu X, Li C. Design and preparation of low-cost α + β titanium alloy based on assessment of Ti-Al-Fe-Cr system. Materials Science and Engineering: A. 2018;**732**:63-69. DOI: 10.1016/j.msea.2018.06.101

[65] Wang H, Wang S, Gao P, Jiang T, Lu X, Li C. Microstructure and mechanical properties of a novel near-α titanium alloy Ti6.0Al4.5Cr1.5Mn. Materials Science and Engineering: A. 2016;**672**:170-174. DOI: 10.1016/j.msea.2016.06.083

[66] Zhu K, Gui N, Jiang T, Zhu M, Lu X, Zhang J, et al. The development of the low-cost titanium alloy containing Cr and Mn alloying elements. Metallurgical and Materials Transactions A. 2014;**45**:1761-1766. DOI: 10.1007/s11661-013-2080-5

[67] Ren L, Xiao W, Chang H, Zhao Y, Ma C, Zhou L. Microstructural tailoring and mechanical properties of a multi-alloyed near β titanium alloy Ti-5321 with various heat treatment. Materials Science and Engineering: A. 2018;**711**:553-561. DOI: 10.1016/j.msea.2017.11.029

[68] Wang H, Xin S, Zhao Y, Zhou W, Zeng W. Plane strain fracture behavior of a new high strength Ti-5Al-3Mo-3V-2Zr-2Cr-1Nb-1Fe alloy during heat treatment. Materials Science and Engineering: A. 2020;**797**:140080. DOI: 10.1016/j.msea.2020.140080

[69] Wu C, Zhao Y, Huang S, Lei L, Zhao Q, Sun Q, et al. Microstructure tailoring and impact toughness of a newly developed high strength Ti-5Al-3Mo-3V-2Cr-2Zr-1Nb-1Fe alloy. Materials Characterization. 2021;**175**:111103. DOI: 10.1016/j.matchar.2021.111103

[70] Wang H, Zhao Q, Xin S, Zhao Y, Zhou W, Zeng W. Microstructural morphology effects on fracture toughness and crack growth behaviors in a high strength titanium alloy. Materials Science and Engineering: A. 2021;**821**:141626. DOI: 10.1016/j.msea.2021.141626

[71] Wang H, Zhao Y, Zhao Q, Xin S, Zhou W, Zeng W. Microstructure evolution and fracture behavior of Ti-5Al-3Mo-3V-2Zr-2Cr-1Nb-1Fe alloy during BASCA heat treatments. Materials Characterization. 2021;**174**:110975. DOI: 10.1016/j.matchar.2021.110975

[72] Wu C, Zhao Q, Huang S, Zhao Y, Lei L, Ren J, et al. Deformation mechanisms in a β-quenched Ti-5321 alloy: In-situ investigation related to slip activity, orientation evolution and stress induced martensite. Journal of Materials Science & Technology. 2022;**112**:36-48. DOI: 10.1016/j.jmst.2021.09.051

[73] Wang H, Xin S, Zhao Y, Zhou W, Zeng W. Forging-microstructure-tensile properties correlation in a new near β high-strength titanium alloy. Rare Metals. 2021;**40**:2109-2117. DOI: 10.1007/s12598-020-01533-y

[74] Takeda O, Ouchi T, Okabe T. Recent progress in titanium extraction and recycling. Metallurgical and Materials Transactions B. 2020;**51**:1315-1328. DOI: 10.1007/s11663-020-01898-6

[75] Xia Y, Zhao J, Tian Q, Guo X. Review of the effect of oxygen on titanium and deoxygenation technologies for

recycling of titanium metal. Journal of Metals. 2019;**71**:3209-3220. DOI: 10.1007/S11837-019-03649-8

[76] Poulsen E, Hall J. Extractive metallurgy of titanium: A review of the state of the art and evolving production techniques. Journal of Metals. 1983;**35**:60-65. DOI: 10.1007/BF03338304

[77] Jiao H, Song W, Chen H, Wang M, Jiao S, Fang D. Sustainable recycling of titanium scraps and purity titanium production via molten salt electrolysis. Journal of Cleaner Production. 2020;**261**:121314. DOI: 10.1016/j.jclepro.2020.121314

[78] Zhang Y, Zhou L, Sun J, Han M, Zhao Y. Evaporation mechanism of aluminum during electron beam cold hearth melting of Ti64 alloy. International Journal of Materials Research. 2009;**100**:248-253. DOI: 10.3139/146.110007

[79] Akhonin S, Trigub N, Zamkov V, Semiatin S. Mathematical modeling of aluminum evaporation during electron-beam cold-hearth melting of Ti-6Al-4V ingots. Metallurgical and Materials Transactions B. 2003;**34**:447-454. DOI: 10.1007/S11663-003-0071-4

[80] Balasundar I, Raghu T, Kashyap B. Modeling the hot working behavior of near-α titanium alloy IMI 834. Progress in Natural Science: Materials International. 2013;**23**:598-607. DOI: 10.1016/j.pnsc.2013.11.004

[81] Zhou Y, Zeng W, Yu H. An investigation of a new near-beta forging process for titanium alloys and its application in aviation components. Materials Science and Engineering: A. 2005;**393**:204-212. DOI: 10.1016/j.msea.2004.10.016

[82] Lutjering G. Influence of processing on microstructure and mechanical properties of (α+β) titanium alloys. Materials Science and Engineering: A. 1998;**243**:32-45. DOI: 10.1016/S0921-5093(97)00778-8

[83] Li C, Huang L, Li C, Hui S, Yu Y, Zhao M, et al. Research progress on hot deformation behavior of high-strength β titanium alloy: Flow behavior and constitutive model. Rare Metals. 2022;**41**:1434-1455. DOI: 10.1007/s12598-021-01861-7

[84] Kumar B, Saxena K, Dey S, Pancholi V, Bhattacharjee A. Processing map-microstructure evolution correlation of hot compressed near alpha titanium alloy (TiHy 600). Journal of Alloys and Compounds. 2017;**691**:906-913. DOI: 10.1016/j.jallcom.2016.08.301

[85] Porntadawit J, Uthaisangsuk V, Choungthong P. Modeling of flow behavior of Ti-6Al-4V alloy at elevated temperatures. Materials Science and Engineering: A. 2014;**599**:212-222. DOI: 10.1016/j.msea.2014.01.064

[86] Peng W, Zeng W, Wang Q, Yu H. Characterization of high-temperature deformation behavior of as-cast Ti60 titanium alloy using processing map. Materials Science and Engineering: A. 2013;**571**:116-122. DOI: 10.1016/j.msea.2013.01.008

[87] Zeng Z, Jonsson S, Zhang Y. Constitutive equations for pure titanium at elevated temperatures. Materials Science and Engineering: A. 2009;**505**:116-119. DOI: 10.1016/j.msea.2008.11.017

[88] Vo P, Jahazi M, Yue S, Bocher P. Flow stress prediction during hot working of near-α titanium alloys. Materials Science and Engineering: A. 2007;**447**:99-110. DOI: 10.1016/j.msea.2006.10.032

[89] Wang K, Zeng W, Zhao Y, Lai Y, Zhou Y. Hot working of Ti-17 titanium

alloy with lamellar starting structure using 3-D processing maps. Journal of Materials Science. 2010;**45**:5883-5891. DOI: 10.1007/S10853-010-4667-1

[90] Kim Y, Song Y, Lee S, Kwon Y. Characterization of the hot deformation behavior and microstructural evolution of Ti-6Al-4V sintered preforms using materials modeling techniques. Journal of Alloys and Compounds. 2016;**676**:15-25. DOI: 10.1016/j.jallcom.2016.03.146

[91] Balasundar I, Ravi K, Raghu T. On the high temperature deformation behaviour of titanium alloy BT3-1. Materials Science and Engineering: A. 2017;**684**:135-145. DOI: 10.1016/j.msea.2016.12.043

[92] Qu Y, Wang M, Lei L, Huang X, Wang L, Qin J, et al. Behavior and modeling of high temperature deformation of an $\alpha + \beta$ titanium alloy. Materials Science and Engineering: A. 2012;**555**:99-105. DOI: 10.1016/j.msea.2012.06.039

[93] Wang G, Hui S, Ye W, Mi X. Hot compressive behavior of Ti-3.0Al-3.7Cr-2.0Fe-0.1B titanium alloy. Transactions of Nonferrous Metals Society of China. 2012;**22**:2965-2971. DOI: 10.1016/S1003-6326(11)61557-4

[94] Rao K, Prasad Y, Suresh K. Hot working behavior and processing map of a γ-TiAl alloy synthesized by powder metallurgy. Materials & Design. 2011;**32**:4874-4881. DOI: 10.1016/j.matdes.2011.06.003

[95] Li C, Narayana P, Reddy N, Choi S, Yeom J, Hong J, et al. Modeling hot deformation behavior of low-cost Ti-2Al-9.2Mo-2Fe beta titanium alloy using a deep neural network. Journal of Materials Science & Technology. 2019;**35**:907-916. DOI: 10.1016/j.jmst.2018.11.018

[96] Luo J, Li M, Yu W. Prediction of flow stress in isothermal compression of Ti-6Al-4V alloy using fuzzy neural network. Materials & Design. 2010;**31**:3078-3083. DOI: 10.1016/j.matdes.2010.01.005

[97] Sun Z, Yang H, Tang Z. Microstructural evolution model of TA15 titanium alloy based on BP neural network method and application in isothermal deformation. Computational Materials Science. 2010;**50**:308-318. DOI: 10.1016/j.commatsci.2010.08.020

[98] Sun Y, Zeng W, Han Y, Ma X, Zhao Y, Guo P, et al. Determination of the influence of processing parameters on the mechanical properties of the Ti-6Al-4V alloy using an artificial neural network. Computational Materials Science. 2012;**60**:239-244. DOI: 10.1016/j.commatsci.2012.03.047

[99] Wu C, Chang H, Wu C, Chen S, Huang S, Huang M, et al. Machine learning recommends affordable new Ti alloy with bone-like modulus. Materials Today. 2020;**34**:41-50. DOI: 10.1016/j.mattod.2019.08.008

[100] Zhang D, Qiu D, Gibson M, Zheng Y, Fraser H, StJohn D, et al. Additive manufacturing of ultrafine-grained high-strength titanium alloys. Nature. 2019;**576**:91-95. DOI: 10.1038/s41586-019-1783-1

[101] Bolzoni L, Ruiz-Navas E, Gordo E. Understanding the properties of low-cost iron-containing powder metallurgy titanium alloys. Materials & Design. 2016;**110**:317-323. DOI: 10.1016/j.matdes.2016.08.010

[102] Weston N, Jackson M. FAST-forge – A new cost-effective hybrid processing route for consolidating titanium powder into near net shape forged components. Journal of Materials Processing

Technology. 2017;**243**:335-346. DOI: 10.1016/j.jmatprotec.2016.12.013

[103] Bolzoni L. Low-cost Fe-bearing powder metallurgy Ti alloys. Metal Powder Report. 2019;**74**:308-313. DOI: 10.1016/j.mprp.2019.01.007

[104] Bolzoni L, Herraiz E, Ruiz-Navas E, Gordo E. Study of the properties of low-cost powder metallurgy titanium alloys by 430 stainless steel addition. Materials & Design. 2014;**60**:628-636. DOI: 10.1016/j.matdes.2014.04.019

[105] Wang J, Liu Y, Rabadia C, Liang S, Sercombe T, Zhang L. Microstructural homogeneity and mechanical behavior of a selective laser melted Ti-35Nb alloy produced from an elemental powder mixture. Journal of Materials Science & Technology. 2021;**61**:221-233. DOI: 10.1016/j.jmst.2020.05.052

[106] Baufeld B, Biest O, Gault R. Additive manufacturing of Ti-6Al-4V components by shaped metal deposition: Microstructure and mechanical properties. Materials & Design. 2010;**31**:S106-S111. DOI: 10.1016/j.matdes.2009.11.032

[107] Lin Z, Song K, Yu X. A review on wire and arc additive manufacturing of titanium alloy. Journal of Manufacturing Processes. 2021;**70**:24-45. DOI: 10.1016/j.jmapro.2021.08.018

[108] Williams S, Martina F, Addison A, Ding J, Pardal G, Colegrove. Wire + arc additive manufacturing. Materials Science and Technology. 2016;**32**:641-647. DOI: 10.1179/1743284715Y.0000000073

[109] Wu B, Pan Z, Ding D, Cuiuri D, Li H, Xu J, et al. A review of the wire arc additive manufacturing of metals: properties, defects and quality improvement. Journal of Manufacturing Processes. 2018;**35**:127-139. DOI: 10.1016/j.jmapro.2018.08.001

[110] Dinovitzer M, Chen X, Laliberte J, Huang X, Frei H. Effect of wire and arc additive manufacturing (WAAM) process parameters on bead geometry and microstructure. Additive Manufacturing. 2019;**26**:138-146. DOI: 10.1016/j.addma.2018.12.013

[111] Busachi A, Erkoyuncu J, Colegrove P, Martina F, Ding J. Designing a WAAM based manufacturing system for defence applications. Procedia CIRP. 2015;**37**:48-53. DOI: 10.1016/j.procir.2015.08.085

[112] Huang R, Riddle M, Graziano D, Warren J, Das S, Nimbalkar S, et al. Energy and emissions saving potential of additive manufacturing: The case of lightweight aircraft components. Journal of Cleaner Production. 2016;**135**:1559-1570. DOI: 10.1016/j.jclepro.2015.04.109

[113] McAndrew A, Rosales M, Colegrove P, Hönnige J, Ho A, Fayolle R, et al. Interpass rolling of Ti-6Al-4V wire+ arc additively manufactured features for microstructural refinement. Additive Manufacturing. 2018;**21**:340-349. DOI: 10.1016/j.addma.2018.03.006

[114] Ali M, Batai S, Sarbassov D. 3D printing: a critical review of current development and future prospects. Rapid Prototyping Journal. 2019;**25**:1108-1126. DOI: 10.1108/RPJ-11-2018-0293

[115] Valluri S, Claremboux V, Kawatra S. Opportunities and challenges in CO_2 utilization. Journal of Environmental Sciences. 2022;**113**:322-344. DOI: 10.1016/j.jes.2021.05.043

[116] Jotzo F, Karplus V, Grubb M, Löschel A, Neuhoff K, Wu L, et al. China's emissions trading takes steps towards big ambitions. Nature Climate Change. 2018;**8**:265-267. DOI: 10.1038/s41558-018-0130-0

[117] Bakhtiari S. Coming out clean: Australian carbon pricing and clean

technology adoption. Ecological Economics. 2018;**154**:238-246. DOI: 10.1016/j.ecolecon.2018.08.004

[118] Murray B, Rivers N. British Columbia's revenue-neutral carbon tax: A review of the latest "grand experiment" in environmental policy. Energy Policy. 2015;**86**:674-683. DOI: 10.1016/j.enpol.2015.08.011

[119] Gao F, Nie Z, Yang D, Sun B, Liu Y, Gong X, et al. Environmental impacts analysis of titanium sponge production using Kroll process in China. Journal of Cleaner Production. 2018;**174**:771-779. DOI: 10.1016/j.jclepro.2017.09.240

[120] Zhang Z, Yu H, Zhang Y, Yang K, Li W, Chen Z, et al. Analysis and optimization of process energy consumption and environmental impact in electrical discharge machining of titanium superalloys. Journal of Cleaner Production. 2018;**198**:833-846. DOI: 10.1016/j.jclepro.2018.07.053

[121] Gohil V, Puri Y. Turning by electrical discharge machining: A review. Proceedings of the Institution of Mechanical Engineers, Part B: Journal of Engineering Manufacture. 2017;**231**:195-208. DOI: 10.1177/0954405415590560

[122] Zhu F, Li L, Cheng X, Ma S, Jiang L, Qiu K. Direct electrochemical reduction of low titanium chlorides into titanium aluminide alloy powders from molten eutectic KCl-LiCl-$MgCl_2$. Electrochimica Acta. 2020;**357**:136867. DOI: 10.1016/j.electacta.2020.136867

[123] Zhu F, Li L, Song W, Zhang D, Ma S, Qiu K. Electrochemical synthesis of Ti-Al-V alloy by chlorination of Ti_2O_3 and V_2O_3 in $AlCl_3$-containing molten chloride salt. Journal of Materials Research and Technology. 2021;**13**:1243-1253. DOI: 10.1016/j.jmrt.2021.05.063

Section 4

Mechanical Properties with Microstructural Considerations

Chapter 5

Shock-Induced Mechanical Response and Microstructure Evolution of Titanium Alloys

Yu Ren

Abstract

The application of titanium alloys in weaponry is increasingly widespread, due to their high specific strength and excellent corrosion resistance. The weapons such as armors must be subjected to intense shock loads caused by explosion and hyper-velocity collision, etc., during service. Therefore, their service performance is closely related to the shock-induced response characteristics of materials, especially the microstructural evolution during the shock pulses and its effect on the mechanical properties. This chapter introduces the research progress on the shock response of some typical titanium alloys such as Ti-6Al-4V, Ti-10V-2Fe-3Al, and Ti-3.5Al-10Mo-8V-1Fe. The effects of alloying composition (alloy type) and stress amplitude on the shock-induced mechanical response and microstructural evolution of titanium alloys are explored through soft recovery shock experiments, quasi-static reloading tests, as well as careful multi-scale microscopic analyses.

Keywords: titanium alloys, alloying composition, shock loading, microstructural evolution, mechanical properties

1. Introduction

Titanium alloys have the advantages of low density, high specific strength, strong corrosion resistance, etc. Presently, titanium materials have been widely used in aerospace, medical, and chemical industries [1–3]. They will undoubtedly play an important role in military fields such as warheads and lightweight armor, due to the many advantages mentioned above [4, 5].

Weapons usually need to withstand extreme loads such as explosions and high-speed collisions during service. At this time, the material often exhibits a very unique behavior, which is different from the response under static loading conditions. Strong shock loads are accompanied by high pressure and high temperature and will cause the material to deform rapidly in a nearly adiabatic state [6]. On the other hand, different from the assumption in statics that the applied force reaches equilibrium in the material instantaneously, the impact load generally propagates in the material in the form of disturbance (wave), and the material state (including microstructures) changes are closely related to the wave propagation process [7]. Therefore, the dynamic behavior of titanium alloys has been the focus of researchers.

As an important topic in the research field of dynamic response of materials, the shock-induced microstructural evolution and its effect on the mechanical behavior of metals have received continuous attention of scholars for nearly 70 years. Although the duration of the shock pulse is short (in general 0.5–2 μs), most of the microstructural evolution processes have enough time to occur and reach an equilibrium state during this period [8]. Therefore, when the shock waves pass through the metals, high-density defects, such as dislocations, twins, will be formed, and phase transformations may also occur [7, 9–13]. The defects and new phases usually increase the strength of materials to some extent, while reducing their plasticity, that is, the so-called shock-induced strengthening effect. However, the strengthening behavior of metallic materials with different crystal structures (such as face-centered cubic, FCC, body-centered cubic, BCC, or hexagonal close-packed, HCP) is very different, and some metals (such as A-70 Ti) even show the phenomenon of "shock softening" under equivalent strain conditions [9, 10]. This suggests that, in addition to defect density, changes in type, morphology, and distribution of defects caused by differences in crystal structure will also have an important influence on the mechanical properties of metallic materials after shock wave loading.

The generation, movement, and interaction of dislocations in metals during shock pulses are affected by many external and intrinsic factors, among which the most important external factor is pressure (shock stress amplitude). The dislocation density increases with increasing shock pressure, which has been recognized by many researchers [7]. However, the dislocation density cannot increase infinitely, but tends to saturate when the pressure exceeds a certain value. Correspondingly, the shock-induced strengthening effect will also be saturated [11]. Another external factor is the shock pulse duration (t_P). In principle, the longer the pulse duration, the closer the shock-induced substructure is to an equilibrium state. However, the time required for dislocation initiation and interaction is in the sub-microsecond order. Therefore, for most metals, if $t_P > 1$ μs, the evolution of the dislocation substructure has sufficient time to reach equilibrium and does not change significantly as the shock loading time increases [12, 13].

The main intrinsic factors that influence the shock-induced microstructure evolution of metals are stacking fault energy (SFE) and Peierls stress (PS) [14, 15]. The Peierls stress (also called Peierls–Nabarro stress) is defined as the minimum shear stress required to move a single dislocation in a perfect crystal at zero temperature, which can be considered as the intrinsic lattice resistance to the dislocation motion [16–18]. The higher the Peierls stress of a metallic material, the less mobility of dislocations in its crystal, which leads to the poor deformability and strain hardening effect [17]. For FCC metals with high SFE (>60 mJ/m^2) and low PS (such as Ni (SFE=128 mJ/m^2 [14])), dislocations are easy to multiply and cross-slip to form mature dislocation cells during shock loading process [13, 14]. Therefore, the shock-induced strengthening effect of this type of metals is very obvious. FCC metals with low SFE (<40 mJ/m^2), such as 304 stainless steel (SFE=21 mJ/m^2 [14, 15]), are more prone to the formation of planar slip, stacking faults, and {111} twinning [19]. On the other hand, although there are more slip systems in BCC metals, their PS is larger than that of FCC metals, so dislocations in BCC metals (such as Ta) proliferate slowly and the mobility is poor. Hence, long-and-straight screw dislocations are often formed and entangled in BCC metals under shock loading conditions. Cell-like structures, dislocation loops, or twins may also be formed [9]. In general, the dislocation density in BCC metals is relatively low after shock loading, which is the main reason for their insignificant strengthening effect.

Compared with FCC and BCC metals, the shock-induced microstructure evolution of HCP metals, such as titanium, magnesium, and their alloys, has not been extensively studied. Koul and Breedis [20] observed a large number of twins in 7 GPa shocked pure Ti and found that there are high-density dislocations in both the twins and the matrix, and the arrangement feature of dislocations is between that of FCC-structured Ni and BCC-structured Fe. Cerreta et al. [10] observed a similar substructure morphology in 11 GPa shocked high-purity Ti. While in the A-70 Ti shocked at the same pressure, planar slip and a small amount of fine twins are the main features of the substructure.

After solution in the β phase region and following rapid quenching, the microstructure of some β titanium alloys consists wholly of equiaxed β grains with BCC structure. If the stability of the β phase is high, the shock-induced substructure is also dominated by dislocations and fine twins [20, 21]. However, if the β phase is in a metastable state, stress/strain-induced phase transformation (SIM) will be activated by shock waves [20].

In addition to the SIM phase transition, titanium and its alloys may also undergo $x \leftrightarrow \omega$ phase transition when the shock wave amplitude is high enough [10, 22–24]. Where x represents α, α′, β or their mixture, and "↔" indicates that the phase transition process is reversible. The shock-induced $x \leftrightarrow \omega$ phase transition has been observed in pure Ti [10] and some titanium alloys such as Ti-64, VT-14, VT-20, and VT-23 [22, 23]. The stress threshold for the $x \leftrightarrow \omega$ phase transition ranges from about 4 GPa to above 35 GPa, which depends on the stress state and the type and content of alloying and interstitial elements [10, 22, 25–27]. Since the ω phase is a brittle phase, dislocations cannot move in it, and this phase mostly exists in the form of dispersed fine particles in the alloy [28, 29], so the $x \rightarrow \omega$ phase transformation may improve the strength of titanium alloys [10].

In this chapter, the research progress on the shock response of some titanium alloys such as extra-low interstitial grade Ti-6Al-4V, Ti-10V-2Fe-3Al and Ti-3.5Al-10Mo-8V-1Fe is presented. The effects of alloying component (alloy type) and stress amplitude on the shock-induced mechanical response and microstructural evolution of titanium alloys are explored through soft recovery shock experiments, quasi-static reloading tests, as well as careful multi-scale microscopic analyses. The research on dynamic behavior of titanium alloy will lay a theoretical foundation for their application in extreme service environment.

2. Materials

In order to explore the influence of content, stability, morphology, and distribution of α and β phase on the high-pressure shock response characteristics of titanium alloys, shock loading experiments of three titanium alloys, one α+β alloy, the famous Ti-6Al-4V (extra-low interstitial grade, Ti-64 ELI [30]), and two metastable β alloys, Ti-10V-2Fe-3Al (Ti-1023) and Ti-3.5Al-10Mo-8V-1Fe (TB3) were carried out. The chemical compositions of three titanium alloys are listed in **Table 1**.

The Ti-64 ELI was tested as received. Ti-1023 and TB3 were solution treated above β-transus temperature (T_β) and then water quenched to room temperature before shock loading. T_β and heat treatment processes of titanium alloys are listed in **Table 2**.

Figure 1 shows the microstructures of three titanium alloys after heat treatments. Ti-64 ELI comprises of α laths (gray area) precipitated from the β matrix (black area), which is called lamellar microstructure, as shown in **Figure 1a**. There is a certain amount of intergranular α phase at the β grain boundaries, indicated by arrows in

Alloys	Alloying elements				Interstitial elements			
	Al	V	Fe	Mo	C	H	O	N
Ti-64 ELI	6.16	3.90	—	—	0.011	0.0022	0.057	0.007
Ti-1023	3.07	10.15	1.68	—	0.007	0.0024	0.075	0.014
TB3	3.62	8.59	1.13	10.13	0.011	0.0028	0.097	0.011

Table 1.
Chemical components of three titanium alloys used (mass fraction, wt.%).

Alloys	Molybdenum equivalent/wt.%	T_β/K	Heat treatment processes
Ti-64 ELI	2.8	1259	As received
Ti-1023	10.6	1087	1123 K/2 h/Water quenched
TB3	18.5	1028	1103 K/25 min/Water quenched

Table 2.
Beta-transus temperature and heat treatment processes for three titanium alloys.

Figure 1a. The average thickness of the α laths is about 2.4 μm. The TEM image of the as-received Ti-64 ELI (**Figure 1b**) shows a low dislocation density before testing. Most of dislocations are short and arranged arbitrarily. The selected area electron diffraction (SAED) pattern (inset in **Figure 1b**) of α phase proves that there is no precipitation of other phases.

Compared with Ti-64 ELI, the two metastable β alloys Ti-1023 and TB3 have higher molybdenum equivalent ($[Mo]_{eq}$), which means the β phase is more stable. After solution treatment, a large amount of β phase can be retained to room temperature. Hence, both Ti-1023 and TB3 are composed of equiaxed β grains (**Figure 1c** and **e**). The average β grain sizes of the two alloys are about 302 and 254 μm, respectively. Because the $[Mo]_{eq}$ of Ti-1023 is at a critical value and the diameter of the bar stock used is large (90 mm), the cooling rate of the core part of the bar stock is slow, so finely dispersed α phases (α_P) are precipitated in the β grains during water quenching, as shown in **Figure 1c**. The average size of these α_P is about 1.3 μm. No α_P precipitated in β grains of TB3 (**Figure 1e**), due to the higher $[Mo]_{eq}$ (higher β phase stability) of this alloy. According to the TEM images (**Figure 1d** and **f**), the dislocation density is also low in both metastable β alloys before shock loading. There is a small amount of fine acicular martensite α″ phase (α''_{th}) activated in Ti-1023 by thermal stress during quenching process (**Figure 1d**). The needle-like α″ phase has an average width of about 41 nm and a length of about 910 nm. The corresponding SAED patterns (inset in **Figure 1d** and **f**) confirm no precipitation of other phases in both Ti-1023 and TB3.

3. Experimental method

3.1 Shock recovery experiments

Effective control of the lateral and rear release waves returning back during unloading process and consequently the residual strain of samples (less than 2%) is the key to correctly evaluate the effect of pre-shock loading on the mechanical properties

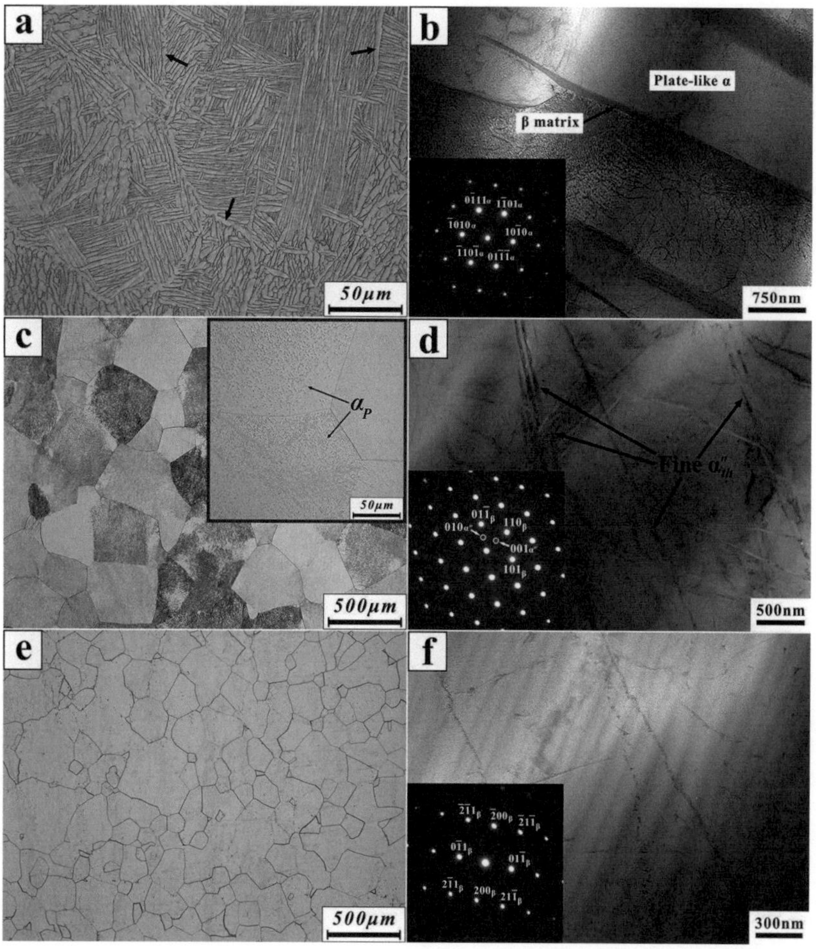

Figure 1.
Microstructures of three titanium alloys before shock loading: (a) and (b) Ti-64 ELI, (c) and (d) Ti-1023, and (e) and (f) TB3. (a), (c), and (e) are optical micrographs. (b), (d) and (f) are TEM images.

and microstructural evolution of materials, which is the so-called soft recovery shock experiment [31]. The size of each component in the soft recovery target assembly, which is closely related to the shock Hugoniot parameters of materials and the experimental parameters such as the shock pressure and the pulse duration, determines the protective effect of the recovery assembly on the recovery sample [31, 32]. The shock pressure (P_H) is set to 5–14 GPa. According to Bourne [8], when the pulse duration is greater than 1 μs, most of the shock-induced microstructural evolution processes have enough time to occur and reach equilibrium. Hence, the pulse duration (t_P) is set to 1.5 μs. The density and shock Hugoniot parameters (C_0 and S) of three titanium alloys used to design the soft recovery target assembly are listed in **Table 3**. C_0 and S are material constants, which are related to the zero-pressure bulk sound velocity and the first pressure derivative of bulk modulus, respectively [35].

In this work, the soft recovery target assembly was designed according to Ref. [31] and illustrated in **Figure 2a**. A 23-mm-diameter and 9-mm-thick Ti-alloy recovery sample was placed behind a 70-mm-diameter and 2.5-mm-thick cover plate. The recovery sample was protected from residual strain and spallation by surrounding an

Alloys	Density ρ_0/g·cm^{-3}	C_0/mm·μs^{-1}	S	References
Ti-64 ELI	4.39	5.05	1.07	[33]
Ti-1023	4.59	4.99	1.06	[34]
TB3	4.82	4.92	1.09	[34]

Table 3.
Shock Hugoniot parameters of three titanium alloys.

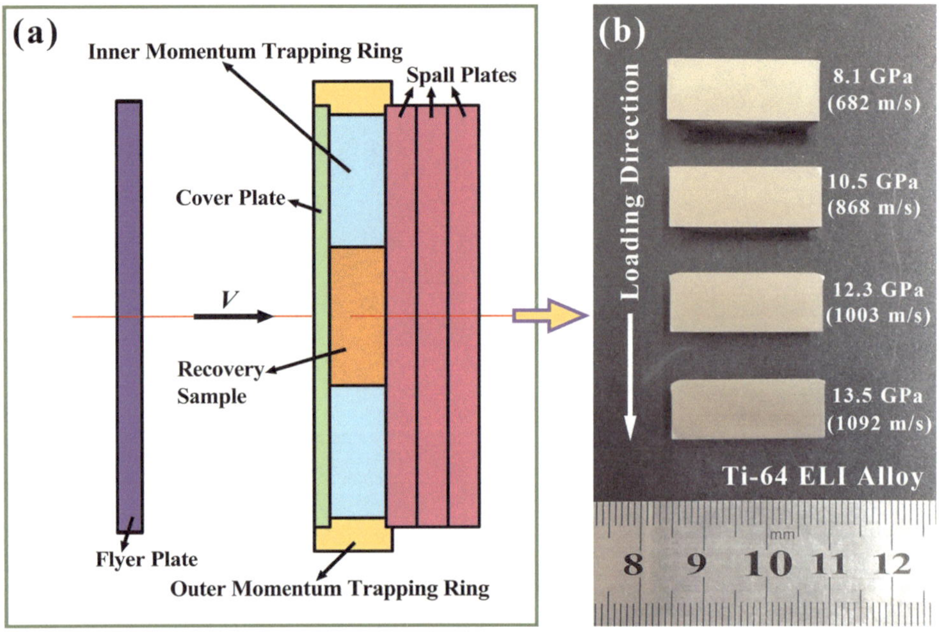

Figure 2.
(a) Schematic diagram of the soft recovery target assembly, and (b) typical macrograph of the longitudinal section of Ti-64 ELI recovery samples shock-loaded at different pressures.

inner momentum trapping ring with outside diameter of 67 mm and supported with three 5-mm-thick spall plates. The inner momentum trapping ring was further concentrically surrounded by an outer momentum trapping ring with an outside diameter of 78 mm. The recovery sample and the inner momentum trapping ring were interference fit together. Two trapping rings, cover plate, and spall plates were assembled with slip fit. Molybdenum disulfide grease was used during assembly of two momentum trapping rings to remove air gaps in the assembly and to facilitate ring separation during recovery.

Using the above-designed soft recovery target assemblies, shock wave loading experiments were carried out on three titanium alloys by a single-stage gas gun with a diameter of 80 mm and a length of 12.5 m. Shock waves with stress amplitudes ranging from ~5 to 14 GPa were yielded in Ti-alloy recovery samples through the impact by flyer plates within the velocity range of 365–1092 ms^{-1}. The thickness of flyer plates ranged from 3.91 to 4.43 mm to ensure a constant pulse duration (t_P) of 1.5 μs. All experiments were conducted in symmetrical impact, which means that the whole shock recovery assembly (containing the recovery sample, two trapping rings, cover plate, and three spall plates) and flyer plate were made of the same material.

The impact surfaces of all recovery assemblies and flyer plates were mechanically polished to ensure that the surface roughness was less than 0.8 μm. The planarity of the target assembly to the flyer plate was controlled by adjusting the specimen mount to better than 1 mrad. The impact velocities were measured by a magnetic induction speed measurement system to an accuracy of about 0.5%.

After shock wave loading, the recovery samples were cut open longitudinally. The longitudinal section of the recovery samples showed that both end faces were flat and no cracks were observed although the edges of the recovery samples deformed slightly, as shown in **Figure 2b**. The residual plastic strain of the shock-loaded recovery samples measured using an outside micrometer is between 0.5 and 1.9%, indicating that the influence of lateral release waves on the recovery samples is deemed to be negligible. The residual strain is defined here as the starting sample thickness minus the recovered sample thickness following shock loading divided by the starting sample thickness.

3.2 Reloading tests and microscopic analyses

Several cylinders (4 mm in diameter and 8 mm in thickness) were sectioned from the center area of the recovery samples for quasi-static reloading compression tests. The reload tests were conducted at 298 K and under a strain rate of 10^{-3} s^{-1} using an MTS 810 hydraulic servo machine. Tests were continuously conducted until the specimens fractured. The reloading direction is parallel to the shock wave loading direction. For comparison, initial materials of Ti-64 ELI, Ti-1023, and TB3 were also compressed under the same conditions. Quasi-static compression tests of all initial and shock prestrained materials were repeated three times to ensure the reliability of the data.

Blocks with dimensions of 7 mm × 5 mm × 2 mm were also cut from the center area of the recovery samples for microstructural analysis. Specimens for optical microscopy (OM) studies were prepared by electrochemically polishing in a solution of 95% ethanoic acid + 5% perchloric acid, followed by etching in solutions of 2% HF + 10% HNO_3 + 88% H_2O (for Ti-64 ELI) and 10% HF + 30% HNO_3 + 60% H_2O (for Ti-1023 and TB3) at room temperature. Then, the blocks were carefully detected using a ZEISS Axio Observer A1m optical microscope. X-ray diffraction (XRD) analyses for phase identification were conducted on a Rigaku Smartlab diffractometer using a Cu-K_α radiation operated at 45 kV and 200 mA. To show the change in the substructure of the preshocked alloy, TEM analyses were conducted using a JEOL JEM-2100 system at an accelerating voltage of 200 kV. For the TEM observations, thin samples with an initial thickness of 0.5 mm were cut from the blocks after OM and XRD examinations, firstly reduced to less than 40 μm thickness by mechanical means, then punched into several standard 3-mm-diameter TEM discs, and finally thinned by ion milling.

4. Mechanical behavior of the postshock titanium alloys

4.1 Stress-strain response

The mechanical behavior of three titanium alloys was detected before and after shock wave loading. Only compression tests were carried out because the tensile specimens were unable to be machined due to the limited size of the recovery samples (23 mm in diameter and 9 mm in thickness). **Figure 3** shows the quasi-static reloading compressive stress-strain curves of the shock prestrained titanium alloys at room

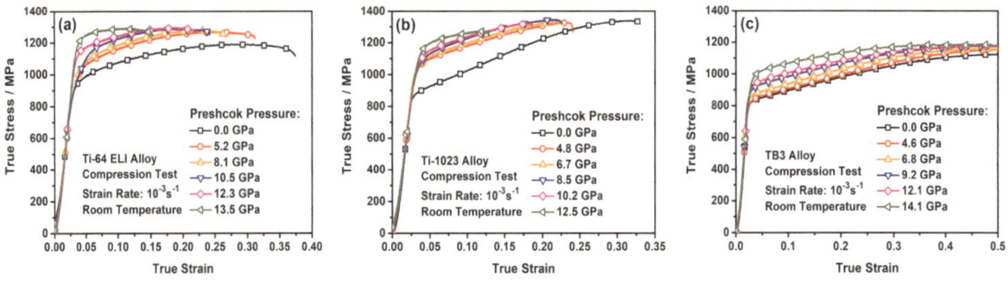

Figure 3.
Typical quasi-static reloading compression true stress-strain curves of three titanium alloys after shocked to different pressures: (a) Ti-64 ELI, (b) Ti-1023, and (c) TB3.

temperature. The curves of shock-free materials are also included in **Figure 3** for comparison. For shock-free (0 GPa) materials, the yield stress of Ti-64 ELI is the highest, followed by Ti-1023, and that of TB3 is the lowest. On the other hand, the plasticity (fracture strain) of TB3 is the best, followed by Ti-64 ELI, and that of Ti-1023 is the worst. After shock wave loading, the shape of the stress-strain curves of three titanium alloys does not change significantly. However, the yield stress and flow stress of the postshock alloys all increase to a certain extent. Besides, the strain hardening rate of the materials decreases gradually with increasing shock pressure (P_H).

To further explore the influence of alloying component and stress amplitude on the shock-induced mechanical behavior, the yield stress (σ_y) and fracture strain (ε_f) of the shock prestrained alloys were plotted versus P_H, as shown in **Figure 4**. The change trend of the reloading σ_y value of Ti-64 ELI with P_H can be roughly divided into two stages, as shown in **Figure 4a**. When the shock pressure is less than 12 GPa, σ_y increases steadily with increasing P_H. When P_H exceeds 12 GPa, the σ_y value of the alloy rises rapidly. As P_H reaches 13.5 GPa, its σ_y value increases by 34% compared with the shock-free material. On the other hand, the plasticity of Ti-64 ELI decreases after shock wave loading. The ε_f value of this alloy decreases in an approximate "parabolic" trend with increasing P_H, as shown in **Figure 4b**.

The change trend of the reloading σ_y value of Ti-1023 with P_H can also be divided into two stages (**Figure 4a**). The σ_y value of this alloy has increased significantly after

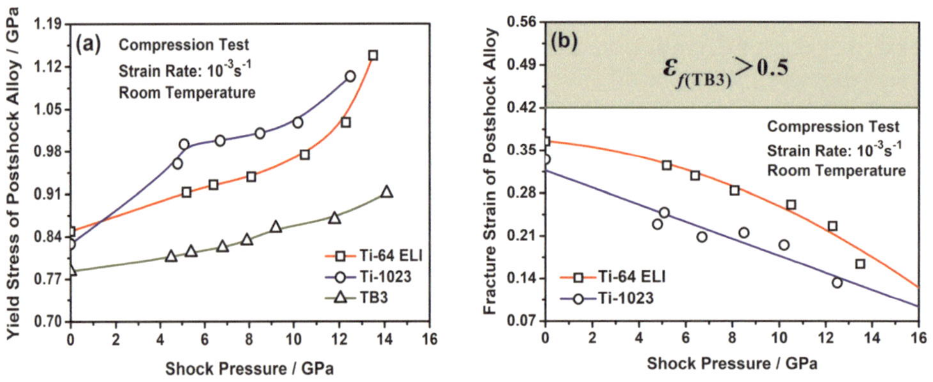

Figure 4.
Quasi-static reloading compression mechanical properties of three titanium alloys after shocked to different pressures: (a) yield stress vs. shock pressure and (b) fracture strain vs. shock pressure.

shock loaded at low amplitudes (~5 GPa). On the other hand, as P_H exceeds 6 GPa, the increasing trend of σ_y slows down with increasing P_H, and the yield strength increases steadily. The plasticity of the shock prestrained Ti-1023 decreases significantly, and the ε_f value of the alloy decreases nearly linearly. Namely, compared with Ti-64 ELI, the shock-induced plasticity deterioration of Ti-1023 is more obvious (**Figure 4b**).

After shock loaded at 4.5–14.1 GPa, the reloading σ_y value of TB3 increases steadily and continuously, as shown in **Figure 4a**. The reloading plastic strain of TB3 is still greater than 0.50 after shocked at different amplitudes (**Figure 4b**), indicating that, when P_H is less than 15 GPa, the shock wave propagation process has no obvious effect on the plasticity of this alloy.

4.2 Hardness

The Vickers hardness (HV) of the postshock titanium alloys was also measured using a WOLPERT 450 SVD Digital Vickers Hardness Tester. The test force and dwell time were set to 5 kgf (49 N) and 10 s, respectively. Generally, the shock prestrain process has no significant effect on the hardness of three titanium alloys, as shown in **Figure 5**. The HV values of Ti-64 ELI and Ti-1023 increase slowly with increasing P_H and tend to be saturated gradually. The HV value of TB3 increases approximately linearly with P_H; however, the increasing trend is not obvious. This indicates that the hardness of titanium alloys is not very sensitive to shock wave loading.

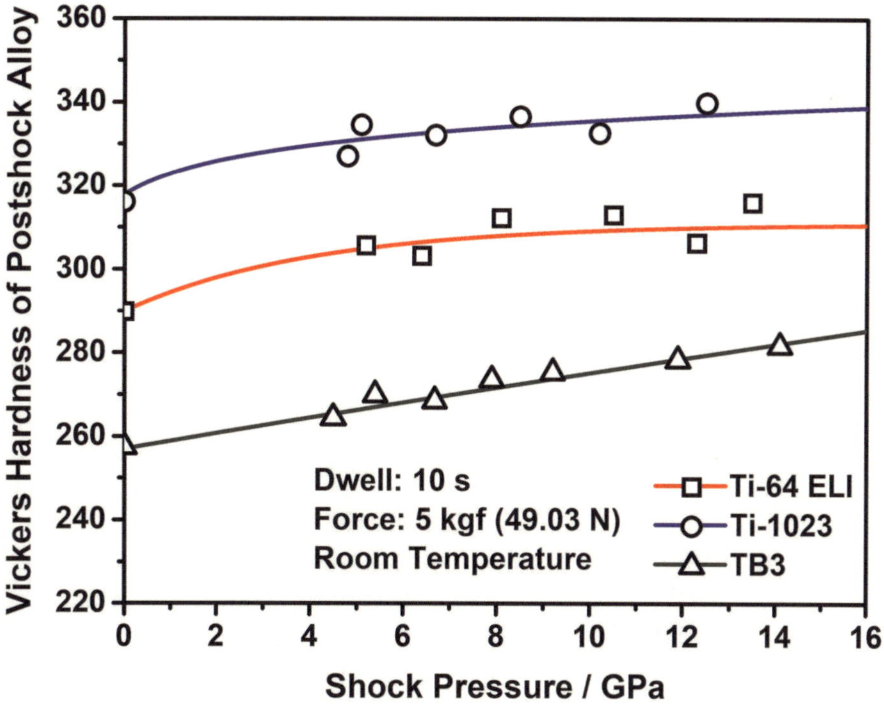

Figure 5.
Vickers hardness-shock pressure curves of three titanium alloys.

5. Shock-induced microstructure evolution

5.1 Phase analysis

Figure 6 shows the typical XRD patterns of the shock-free and postshock titanium alloys. All spectral lines were normalized. Compared with the initial material, except for peaks of α and β phases, no new peaks were detected in the slow-scanning XRD patterns of the postshock Ti-64 ELI, indicating that, when P_H is less than 14 GPa, no new phase is generated in the alloy during the shock pulse.

For the metastable β alloy, Ti-1023, in addition to the peaks involving α_P and β phases, there occur new peaks in the XRD spectrum of the alloy shocked at 4.8 GPa. These new peaks correspond to the α″ phase [36, 37], that is, the shock-induced martensite (SIM). It suggests that the β→α″ phase transformation has already occurred in Ti-1023 even if the shock pressure is relatively low. As P_H increases, the peaks of α″ phase always exist, and no new peaks appear.

For another metastable β alloy, TB3, when the shock pressure is relatively low (4.5 GPa), except for the β peaks, no new peaks appear in the XRD pattern. However, at a higher shock pressure, the α″ peaks were observed. This result indicates that the stability of β phase in TB3 is higher than that in Ti-1023.

5.2 Metallographic microstructure

Figure 7 presents the optical microstructures of Ti-64 ELI, Ti-1023, and TB3 preshocked at different pressures. For the α+β alloy, Ti-64 ELI, compared with the

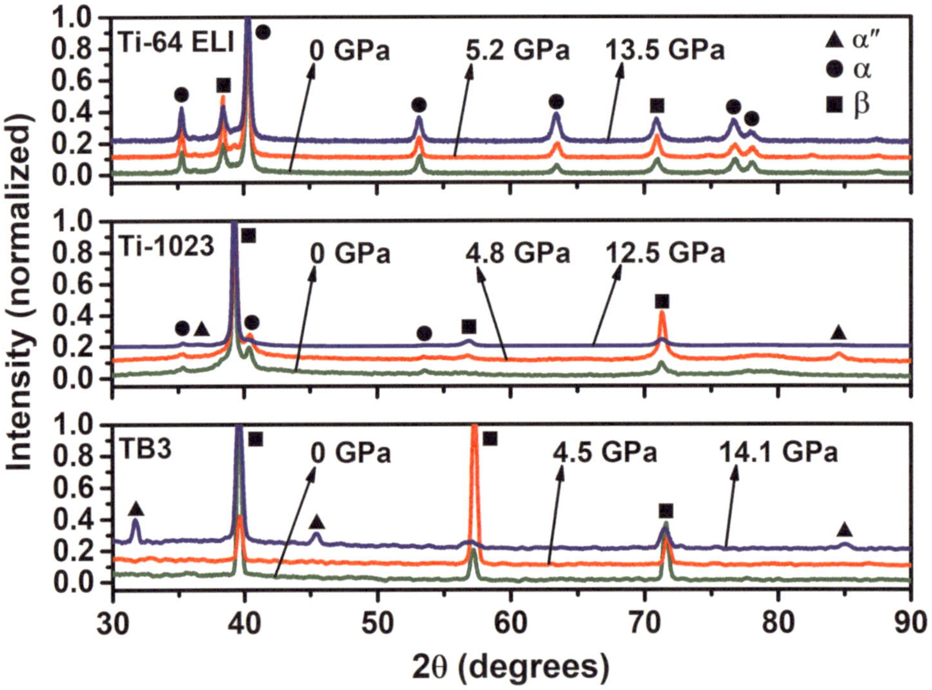

Figure 6.
Slow X-ray scans from titanium alloys after shocked to different pressures.

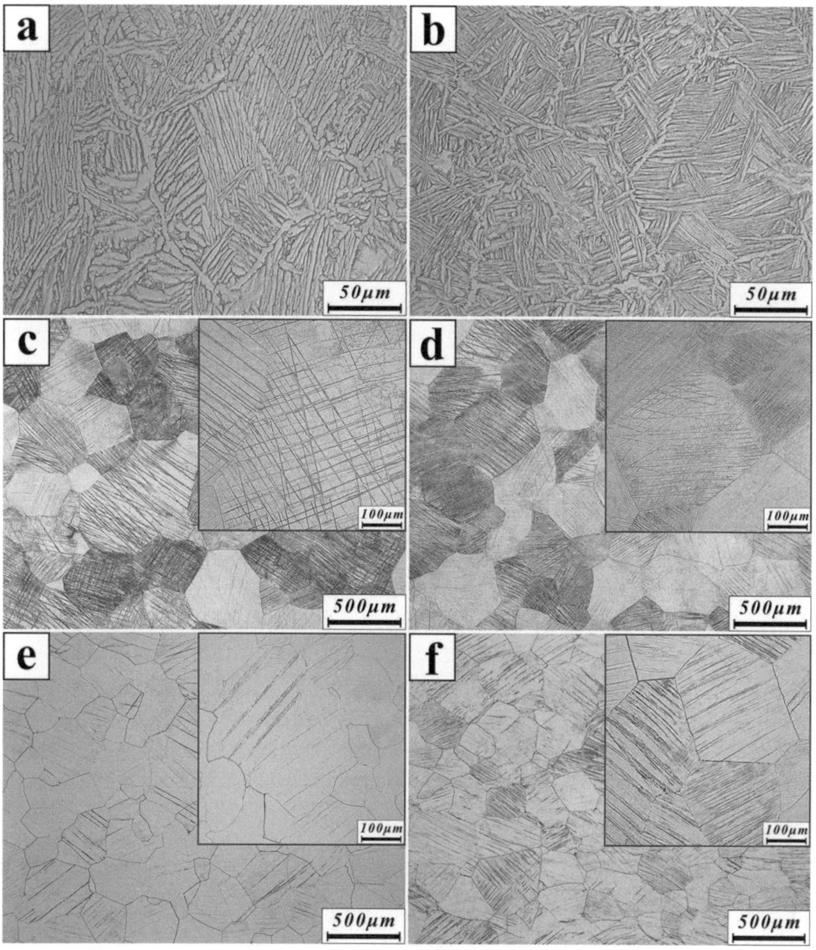

Figure 7.
Optical microstructures of the postshock titanium alloys: (a) Ti-64 ELI shocked at 5.2 GPa, (b) Ti-64 ELI shocked at 13.5 GPa, (c) Ti-1023 shocked at 4.8 GPa, (d) Ti-1023 shocked at 12.5 GPa, (e) TB3 shocked at 4.5 GPa, and (f) TB3 shocked at 14.1 GPa.

initial material (**Figure 1a**), the metallographic microstructure of the alloy does not change, even if P_H reaches 13.5 GPa (**Figure 7a** and **b**). For Ti-1023, after shocked at 4.8 GPa, the structure of the alloy has significantly changed: A large number of needle-like phases are produced in the equiaxed β grains (**Figure 7c**). According to the XRD analysis mentioned in Section 5.1, these acicular phases are SIMs. Most α″ laths fully develop to go throughout the entire β grains and intersect each other. As P_H increases, the morphology of α″ has no obvious change, but its amount (density) slightly increases (**Figure 7d**). For TB3 with higher $[Mo]_{eq}$, the number of needle-like α″ phases is very limited at relatively low pressure (4.5 GPa), as shown in **Figure 7e**. Hence, they cannot be detected by XRD (**Figure 6**). The density of SIM phase increases gradually with increasing P_H. Besides, at the same shock pressure, the amount of α″ phase in TB3 is not only less than that in Ti-1023 but most of the α″ laths in TB3 are arranged parallel to each other and rarely staggered (**Figure 7f**).

5.3 Substructure evolution

Figure 8 shows the TEM images of the substructures in the α phase of the Ti-64 ELI shock loaded at different pressures. Compared with the initial material (**Figure 1b**), after shocked at a relatively low pressure (5.2 GPa), the dislocations in the α plates become longer. The density and entanglement degree of the dislocations are increased. Some small dislocation cusps on the long dislocations (indicated by the arrows in **Figure 8a**) were also observed. When P_H increases to 8.1 GPa (**Figure 8b**), the dislocations in the α phase proliferate greatly, forming a substructure dominated by net-like planar slip. This dislocation substructure is very similar to that in the quasi-statically deformed commercial Ti-64 [38]. However, the distribution of planar slip dislocations formed by shock wave loading is more uniform, indicating that the shock-induced instantaneous deformation of Ti alloys is more uniform. When the shock pressure reaches 10.5 GPa (**Figure 8c**), the dislocation density continues to increase, and the dislocation entanglement is more serious. At this time, the dislocation substructure has lost the regular arrangement characteristic and tended to form a band/cluster structure. With the further increase of P_H to 12.3 and 13.5 GPa (**Figure 8d** and **f**), the high-density dislocations in the α phase are further entangled with each other, and dislocation clusters or slip bands are finally formed. The SAED pattern obtained from the α region in the preshocked Ti-64 ELI shows the existence only of α phase, consistent to the XRD analysis results.

In addition to the entangled high-density dislocations, fine twins were also observed in the Ti-64 ELI after shocked at higher pressures (>12 GPa), as shown in **Figure 8d–f**. When P_H is 12.3 GPa, the number of twin laths is still small, and the average thickness of these twins is about 309 nm (**Figure 8d**). The SAED pattern (**Figure 8e**) confirms the twins are $\{10\bar{1}2\}$ twins [39]. Besides, similar to magnesium

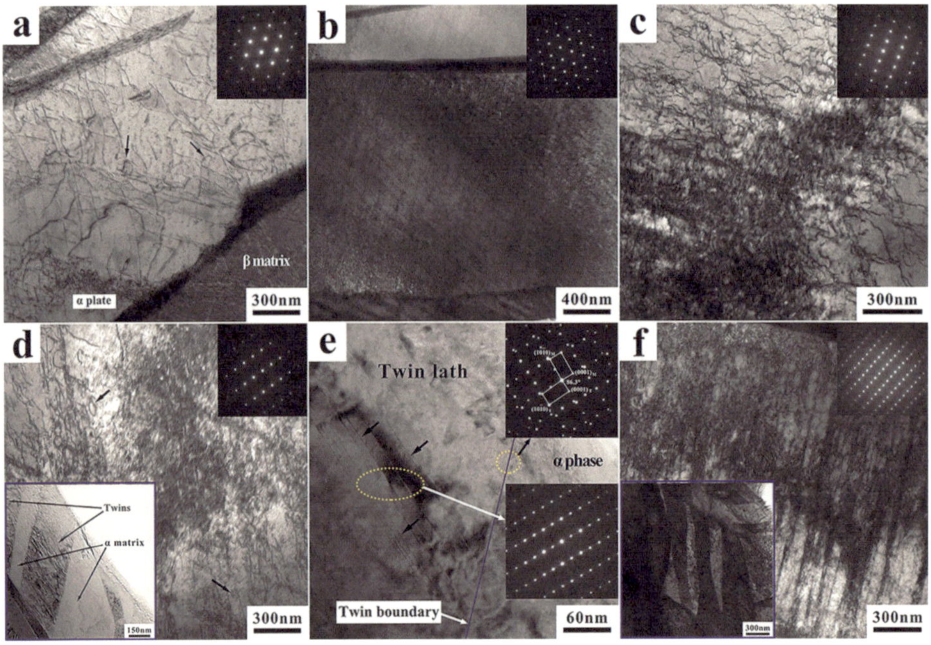

Figure 8.
Bright field TEM images showing dislocation and twin substructures in the α phase within Ti-64 ELI shock loaded to different pressures: (a) 5.2 GPa, (b) 8.1 GPa, (c) 10.5 GPa, (d) and (e) 12.3 GPa, and (f) 13.5 GPa.

alloys [40], stacking faults (SFs) were also observed inside the twin laths (indicated by arrows in **Figure 8e**). When P_H further increases to 13.5 GPa, the number of twins in Ti-64 ELI increases substantially (**Figure 8f**). The average thickness of the twin laths is about 230 nm, which is slightly smaller than that of twins generated at 12.3 GPa.

Figure 9 presents the α″ substructure characteristics in the equiaxed β grains of the preshocked Ti-1023. According to these TEM images, one can find that the needle-like α″ phases observed in the optical micrographs (**Figure 7c** and **d**) are actually martensitic lath bundles. These SIM bundles are made up of many finer interweaving martensites. The SAED patterns in **Figure 9** also reflect the complex internal structure of the α″ bundles. The bundles have a thickness of about 1.58 to 2.51 μm, and the average thickness is about 2.04 μm. As P_H increases, the thickness of the α″ bundle decreases, but its internal substructures become more regular. This implies the rotation and coalescence of α″ laths during the shock pulses with higher stress amplitudes. Besides, a large number of interlaced fine martensites were also observed, as shown in **Figure 9b** and **e**.

Figure 10 shows the substructural features in the TB3 shocked at different pressures. Compared with the initial material (**Figure 1f**), after shock loaded at a relatively low pressure (4.5 GPa), dislocation walls consisting of many short dislocations arranged in parallel are formed in the β grains, as shown in **Figure 10a**. The width of these dislocation walls is about 375 nm. When P_H continues to increase, the dislocation density gradually increases, which leads to an increase in the number of dislocation walls and intersection between them (**Figure 10b** and **c**). The width of the dislocation walls gradually decreases with increasing P_H. When P_H exceeds 10 GPa (**Figure 10d** and **e**), the dislocations in the β phase are severely entangled, and the dislocation walls begin to transform into slip bands, which are similar to those formed

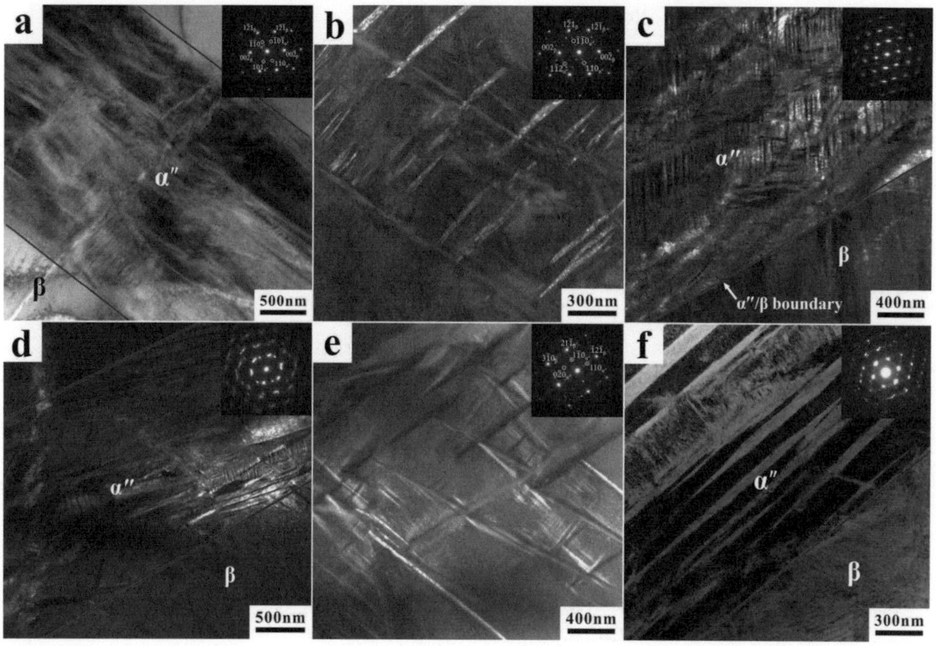

Figure 9.
TEM images showing α″ substructures in the β phase within Ti-1023 shock loaded to different pressures: (a) and (b) 4.8 GPa, (c) 8.5 GPa, (d) and (e) 10.2 GPa, and (f) 12.5 GPa. (b)–(d) are dark field TEM images.

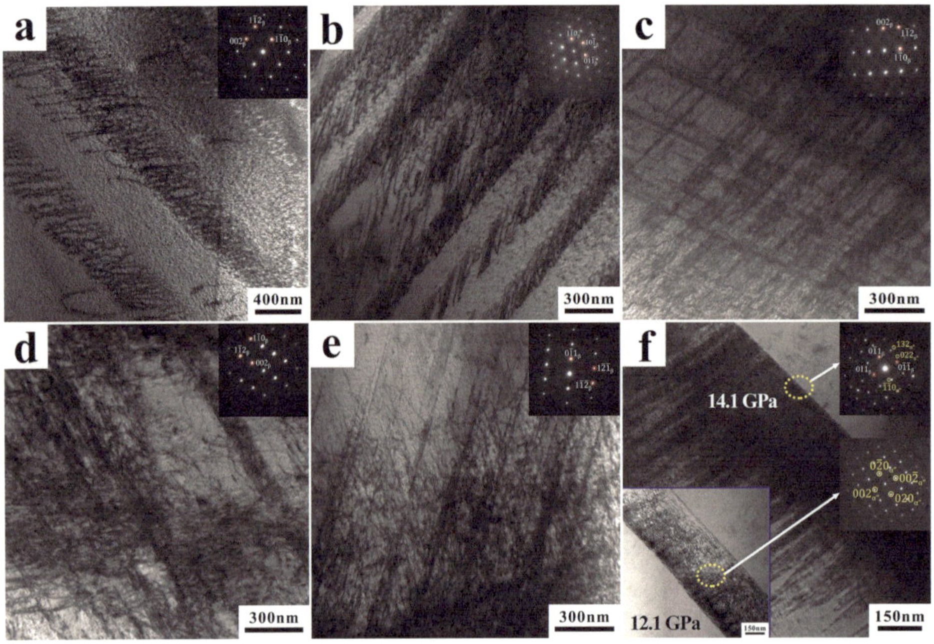

Figure 10.
Bright field TEM images showing dislocation and α substructures in the β phase within TB3 shock loaded to different pressures: (a) 4.5 GPa, (b) 6.8 GPa, (c) 9.2 GPa, (d) 12.1 GPa, (e) 14.1 GPa, and (f) 12.1 and 14.1 GPa.

in 13.5 GPa-shocked Ti-64 ELI. Besides, no additional spots appear in the SAED patterns of the postshock alloys, suggesting that, when the shock pressure is less than about 15 GPa, the β→ω phase transition does not occur in TB3.

The substructure of the α″ lath in TB3 is presented in **Figure 10f**. Compared with that in Ti-1023, the internal structure of the α″ lath in TB3 is relatively simple. The α″ laths have a thickness of about 0.71–1.57 μm, which is narrower than the SIM bundle formed in Ti-1023.

6. Discussion

The shock preloading and quasi-static reloading results indicate that the propagation of shock waves can significantly change the microstructure and further the mechanical properties of titanium alloys. For Ti-64 ELI, as the shock pressure is lower than 12 GPa, the shock-induced substructure of this alloy is dominated by dislocations, similar to those dislocation substructures generated in A-70 Ti [10] and commercial Ti-64 [38]. The shock-induced high-density dislocations (clusters or bands) prevent further proliferation and movement of dislocations in the material during subsequent reloading deformation process, resulting in an increase in the yield strength of Ti-64 ELI (**Figure 4a**). No dislocation cells or cell-like structures were observed, due to the low symmetry and relatively high Peierls stress of the α phase with HCP crystal structure. Therefore, when shock loaded at relatively low pressures (<12 GPa), Ti-64 ELI does not show a very obvious shock-induced strengthening effect like Ni [13, 14]. However, as the shock pressure exceeds 12 GPa, twinning is activated in α phase. The twin laths are able to split α grains, resulting in grain refinement. Hence, as a new strengthening mechanism, twinning makes the reloading yield strength of Ti-64 ELI rapidly increase, showing a more obvious shock strengthening effect (**Figure 4a**). The generation of a large number of micro defects during the shock pulse hinders the movement of dislocations, leading to the dislocation pile-up more serious and stress concentration in the process of reloading deformation. This makes it easier for micro damages to nucleate and expand and finally results in the decline in the plasticity of the postshock Ti-64 ELI, as shown in **Figure 4b**.

For Ti-1023 with low β phase stability, the SIM phase transformation has occurred at 4.8 GPa. A large number of interlaced α″ laths were formed and went throughout the entire β grains. The α″ phase possesses the orthorhombic crystal structure [41], and it has fewer slip systems than that in the matrix β phase with BCC structure [36]. Therefore, compared with the β phase, the α″ phase is more difficult to coordinate the plastic strain. Moreover, the martensite laths are mostly interlaced with each other and their internal structure is very complex (**Figures** 7 and **9**). These substructures will greatly hinder the dislocation movement and cause the interaction between the dislocations and the martensites or cause the dislocations pile-up at the α″/β boundaries. Consequently, the deformation resistance and yield stress of the postshock material dramatically increase during the reloading compression process even if the shock pressure is relatively low (4.8 GPa), as shown in **Figure 4a**. The similar strengthening effect caused by SIM transition was also observed in the 7 GPa shocked Ti-12 Mo [20]. However, the number of α″ phases in Ti-1023 has tended to be saturated at 4.8 GPa, and its amount has little change with increasing P_{H}. Hence, the increase in yield stress of the postshock alloy tends to be gentle (**Figure 4a**). Compared with dislocation clusters/bands and twins, the interlaced acicular α″ martensites hinder the movement of dislocations more seriously, so the plasticity of Ti-1023 decreases sharply after shock wave loading.

When the shock pressure amplitude is similar, the amount of α″ martensite formed in TB3 is far less than that formed in Ti-1023 (**Figure** 7), due to the higher stability of β phase in TB3. Most of the α″ laths in TB3 are arranged in parallel and rarely interlaced. Their internal structure is also relatively simple (**Figure 10f**). The smaller grain size also inhibits the occurrence of β→α″ phase transition to some extent [42]. Therefore, the shock-induced α″ phase has little contribution to the shock strengthening effect of TB3; however, it also has little effect on the plasticity of the alloy (**Figure 4b**). On the other hand, although a certain amount of dislocation walls or slip bands are formed in TB3, the degree of dislocation proliferation and entanglement is not as good as that in Ti-64 ELI under shock loading conditions. Besides, no twinning occurred in TB3. However, shock-induced twins were observed in some other β titanium alloys such as Ti-26 [20]. The higher β phase stability (higher $[\mathrm{Mo}]_{eq}$) of Ti-26 alloy may completely suppress the occurrence of SIM transition. Hence, twinning replaces SIM transformation to coordinate the deformation during shock loading pulse. The single type of micro defect (only dislocations) and the low density of dislocations and α″ phases make the shock-induced strengthening effect in TB3 limited, and the yield stress increases gently, as shown in **Figure 4a**.

The extent of the shock-induced strengthening in titanium alloys significantly depends on the type, morphology, and density of the defects that are decided by the alloying composition (alloy type). For α+β alloy, Ti-64 ELI, the stability of both α and β phases is very high due to the enrichment of α- and β-stabilizing elements in α and β phases, respectively. Hence, the shock-induced substructure evolution in Ti-64 ELI is dominated by dislocation multiplication and interaction. Twinning is activated as dislocation proliferation tends to saturate during shocked at higher pressures (>12 GPa). For metastable β alloys, Ti-1023 and TB3, the β-stabilizing elements are uniformly distributed in the β phase of the alloys after solution treatment, and the β phase is in an unstable state. Therefore, except for dislocation substructures, shock-induced α″ martensites are also generated. However, the number, morphology, and distribution of the α″ phase depend on the stability of the β phase, which is determined by the content of β-stabilizing elements. The content of β-stabilizing elements in TB3 is higher than that in Ti-1023; hence, the proliferation, growth, and interlacing of α″ martensites are weaker in TB3 due to the higher stability of β phase in this alloy.

In general, when P_H is less than 15 GPa, with the increase of $[Mo]_{eq}$ (from Ti-64 ELI to TB3), the shock-induced microstructure evolution characteristics of titanium alloys can be roughly summarized as follows: planar slip (tangled dislocations, slip bands) + twins (Ti-64 ELI) → interlaced α″ martensites (Ti-1023) → paralleled α″ martensites + planar slip (dislocation walls, slip bands) (TB3). That is to say, the main deformation mechanism of the three titanium alloys during the shock pulse undergoes the evolution of "slip/twinning → β/α″ phase transition → β/α″ phase transition + slip."

7. Conclusions

In this chapter, the effects of alloying composition and stress amplitude on the shock-induced mechanical behavior and microstructure evolution of three titanium alloys are revealed through a series of soft recovery shock experiments, quasi-static reloading tests, and multi-scale microscopic analyses. The major conclusions are as follows:

1. When P_H is less than 12 GPa, the shock-induced substructure of Ti-64 ELI is dominated by tangled dislocations. When P_H exceeds 12 GPa, dislocation entanglement is further intensified to form dislocation clusters/bands, and $\{10\bar{1}2\}$ twin is also activated. Correspondingly, the yield stress of the postshock alloy first increases slowly and then increases rapidly.

2. A large amount of α″ martensite phase is produced in Ti-1023 during the shock pulses. Most of the α″ laths have a complex internal structure, interlace with each other, and go through the entire β grains. The existence of high-density martensite laths makes Ti-1023 show a significant shock strengthening effect even if the shock pressure is relatively low. However, the plasticity of the preshocked alloy is also degraded seriously.

3. Higher β phase stability and smaller grain size inhibit the β→α″ transition in TB3. The density of the martensite phase is low in this alloy, and the α″ laths are mostly arranged in parallel with few intersections. The internal structure of the α″ laths is relatively simple. A certain amount of dislocation walls or slip bands also form in TB3, but no twins generate during the shock loading process. Hence, the shock-induced strengthening effect in TB3 is limited and the yield stress increases gently due to the single type of micro defect and the low density of defects and α″ phases.

4. When P_H is less than 15 GPa, with the increase of $[Mo]_{eq}$ (from Ti-64 ELI to TB3), the main deformation mechanism of the three titanium alloys during

shock pulse undergoes the evolution of "slip/twinning → β/α″ phase transition → β/α″ phase transition + slip."

Acknowledgements

This study was funded by China Postdoctoral Science Foundation (Grant number: 2021M690012), the Young Scientists Fund of the National Natural Science Foundation of China (Grant number: 51501064), and the Fundamental Research Funds for the Central Universities (Grant number: 2019MS012).

Conflict of interest

The authors declare no conflict of interest.

Author details

Yu Ren[1,2]

1 School of Mechatronical Engineering, Beijing Institute of Technology, Beijing, China

2 School of energy, power and mechanical engineering, North China Electric Power University, Beijing, China

*Address all correspondence to: renyu@bit.edu.cn, renyuljz@163.com

References

[1] Gurrappa. Characterization of titanium alloy Ti-6Al-4V for chemical, marine and industrial applications. Materials Characterization. 2003;**51**(2–3):131-139

[2] Zhao Q, Sun Q, Xin S, Chen Y, Wu C, Wang H, et al. High-strength titanium alloys for aerospace engineering applications: A review on melting-forging process. Materials Science and Engineering A. 2022;**845**:143260

[3] Guo Amy XY, Cheng L, Zhan S, Zhang S, Xiong W, Wang Z, et al. Biomedical applications of the powder-based 3D printed titanium alloys: A review. Journal of Materials Science and Technology. 2022;**125**:252-264

[4] Millett JCF, Whiteman G, Bourne NK, Gray GT III. The role of anisotropy in the response of the titanium alloy Ti–6Al–4V to shock loading. Journal of Applied Physics. 2008;**104**:073531

[5] Yang Y, Jiang Z, Wang C, Hu HB, Tang TG, Zhang HS, et al. Effects of the phase interface on initial spallation damage nucleation and evolution in dual phase titanium alloy. Materials Science and Engineering A. 2018;**731**:385-393

[6] Davison L. Fundamentals of Shock Wave Propagation in Solids. Berlin Heidelberg: Springer-Verlag; 2008

[7] Meyers MA. Dynamic Behavior of Materials. New York: John Wiley & Sons, Inc.; 1994

[8] Bourne NK. On the failure and dynamic performance of materials. Experimental Mechanics. 2012;**52**(2): 153-159

[9] Gray GT III, Vecchio KS. Influence of peak pressure and temperature on the structure/property response of shock-loaded Ta and Ta–10W. Metallurgical and Materials Transactions A. 1995;**26A**(10):2555-2563

[10] Cerreta E, Gray GT III, Lawson AC, Mason TA, Morris CE. The influence of oxygen content on the α to ω phase transformation and shock hardening of titanium. Journal of Applied Physics. 2006;**100**:013530

[11] Champion AR, Rohde RW. Hugoniot equation of state and the effect of shock stress amplitude and duration on the hardness of Hadfield steel. Journal of Applied Physics. 1970;**41**(5):2213-2223

[12] Wongwiwat K, Murr LE. Effect of shock pressure, pulse duration, and grain size on shock-deformation twining in molybdenum. Materials Science and Engineering. 1978;**35**(2):273-285

[13] Murr LE, Kuhlmann-Wilsdorf D. Experimental and theoretical observations on the relationship between dislocation cell size, dislocation density, residual hardness, peak pressure and pulse duration in shock-loaded nickel. Acta Metallurgica. 1978;**26**(5):847-857

[14] Murr LE. Residual microstructure-mechanical property relationships in shock-loaded metals and alloys. In: Meyers MA, Murr LE, editors. Shock Waves and High-Strain-Rate Phenomena in Metals. New York: Plenum Press; 1981. pp. 607-673

[15] Bourne NK, Millett JCF, Gray GT III. On the shock compression of polycrystalline metals. Journal of Materials Science. 2009;**44**(13):3319-3343

[16] Nabarro FRN. Fifty-year study of the Peierls-Nabarro stress. Materials Science and Engineering A. 1997;**234–236**:67-76

[17] Edagawa K, Kamimura Y, Iskandarov AM, Umeno Y, Takeuchi S. Peierls stresses estimated by a discretized Peierls–Nabarro model for a variety of crystals. Materialia. 2019;**5**: 100218

[18] Chen H, Levitas VI, Xiong L, Zhang X. Stationary dislocation motion at stresses significantly below the Peierls stress: Example of shuffle screw and 60° dislocations in silicon. Acta Materialia. 2021;**206**:116623

[19] Murr LE, Staudhammer KP. Effect of stress amplitude and stress duration on twinning and phase transformations in shcok-loaded and cold-rolled 304 stainless steel. Materials Science and Engineering. 1975;**20**(1):35-46

[20] Koul MK, Breedis JF. Strengthening of titanium alloys by shock deformation. In: Jaffee RI, Promisel NE, editors. The Science, Technology and Application of Titanium. Oxford: Pergamon Press; 1970. pp. 817-828

[21] Rack HJ. Residual strength of shock loaded RMI 38644. Metallurgical Transactions A. 1976;**7A**(10):1571-1576

[22] Mescheryakov YI, Divakov AK, Zhigacheva NI. Shock-induced phase transformation and vortex instabilities in shock loaded titanium alloys. Shock Waves. 2000;**10**:43-56

[23] Mescheryakov YI, Divakov AK. Affect of shock-induced phase transformations on dynamic strength of titanium alloys. International Journal of Impact Engineering. 2001;**26**: 497-508

[24] Divakov AK, Mescheryakov YI, Zhigacheva NI, Barakhtin BK, Gooch WA. Spall strength of titanium alloys. Physical Mesomechanics. 2010;**13** (3–4):113-123

[25] Greeff CW, Trinkle DR, Albers RC. Shock induced α-ω transition in titanium. Journal of Applied Physics. 2001;**90**(5):2221-2226

[26] Errandonea D, Meng Y, Somayazulu M, Häusermann D. Pressure-induced α→ω transition in titanium metal: A systematic study of the effects of uniaxial stress. Physica B: Condensed Matter. 2005;**355**(1-4): 116-125

[27] Hennig RG, Trinkle DR, Bouchet J, Srinivasan SG, Albers RC, Wilkins JW. Impurities block the α to ω martensitic transformation in titanium. Nature Materials. 2005;**4**:129-133

[28] Brotzen FR, Harmon EL, Troiano AR. Decomposition of beta titanium. Journal of Metals. 1955;**7**: 413-419

[29] Hickman BS. The formation of omega phase in titanium and zirconium alloys: A review. Journal of Materials Science. 1969;**4**:554-563

[30] ASTM B348-11. Standard Specification for Titanium and Titanium Alloy Bars and Billets, Book of ASTM Standards, vol. 02.04

[31] Gray GT III. Influence of shock-wave deformation on the structure/property behavior of materials. In: Asay JR, Shahinpoor M, editors. High Pressure Shock Compression of Solids. New York: Springer-Verlag; 1993. pp. 187-215

[32] Bourne NK, Gray GT III. Computational design of recovery experiments for ductile metals. Proceedings of the Royal Society A. 2005;**461**:3297-3312

[33] Ren Y, Wang F, Tan C, Wang S, Yu X, Jiang J, et al. Shock-induced

mechanical response and spall fracture behavior of an extra-low interstitial grade Ti-6Al-4V alloy. Materials Science and Engineering A. 2013;**578**:247-255

[34] Ren Y, Wang F, Tan C, Wang S, Yu X, Jiang J, et al. Effect of shock-induced martensite transformation on the postshock mechanical response of metastable β titanium alloys. Journal of Alloys and Compounds. 2013;**578**:547-552

[35] McQueen RG, Marsh SP. Equation of state for nineteen metallic elements from shock-wave measurements to two megabars. Journal of Applied Physics. 1960;**31**(7):1253-1269

[36] Bhattacharjee A, Varma VK, Kamat SV, Gogia AK, Bhargava S. Influence of β grain size on tensile behavior and ductile fracture toughness of titanium alloy Ti-10V-2Fe-3Al. Metallurgical and Materials Transactions. 2006;**37A**:1423-1433

[37] Ramarolahy A, Castany P, Prima F, Laheurte P, Péron I, Gloriant T. Microstructure and mechanical behavior of superelastic Ti-24Nb-0.5O and Ti-24Nb-0.5N biomedical alloys. Journal of the Mechanical Behavior of Biomedical Materials. 2012;**9**:83-90

[38] Follansbee PS, Gray GT III. An analysis of the low temperature, low and high strain-rate deformation of Ti-6Al-4V. Metallurgical Transactions A. 1989; **20A**(5):863-874

[39] Mahajan S, Williams DF. Deformation twinning in metals and alloys. International Metallurgical Reviews. 1973;**18**:43-61

[40] Zhou B, Sui M. High density stacking faults of $\{10\bar{1}1\}$ compression twin in magnesium alloys. Journal of Materials Science & Technology. 2019; **35**(10):2263-2268

[41] Duerig TW, Albrecht J, Richter D, Fischer P. Formation and reversion of stress induced martensite in Ti-10V-2Fe-3Al. Acta Metallurgica. 1982;**30**(12): 2161-2172

[42] Paradkar A, Kamat SV, Gogia AK, Kashyap BP. Trigger stress for stress-induced martensitic transformation during tensile deformation in Ti-Al-Nb alloys: Effect of grain size. Metallurgical and Materials Transactions A. 2008;**39A**(3):551-558

Chapter 6

Welding Properties of Titanium Alloys Grade 5

Alireza Farrokhi

Abstract

Titanium alloys have attractive properties that have been widely used in various fields due to these properties. The biocompatibility of titanium has caused its usages in the pharmaceutical industry. Its good corrosion resistance has resulted in its many applications in petrochemical and marine industries. Meanwhile, the most important and major application of titanium and its alloys can be found in the aerospace industry in a variety of spacecraft, space rockets, and satellites. The most important reason for the widespread use of these alloys in the aerospace industry is the high ratio of strength to their weights. However, the most important factor limiting the usage of titanium and its alloys is its high price. In this chapter, we first introduced titanium and its alloys. Then, according to mainstream of the design, titanium alloy bonding methods and welding problems have been investigated. The purpose of this study is to investigate the processes of arc welding of tungsten electrode-neutral gas with different alloy heat inputs for a titanium alloy of Ti-6Al-4 V and to provide comprehensive welding instructions. So, after introducing the pure commercial titanium properties, we examined the properties, specifications and welding of Ti-6Al-4 V. Given the focus on the mentioned process, the generalities of this welding process have been described.

Keywords: procedure qualification record, welding procedure specification, Ti-6 Al-4 V titanium alloy grade 5, gas tungsten inert gas, tungsten inert gas, fusion zone, heat affected zone

1. Introduction

High ratio of strength to weight and the desired properties of corrosion resistance of titanium alloys have provided special applications to these alloys. However, high actual cost of making these alloys has limited their usages. The price of this metal is considerably higher than widely used industrial metals such as steel, copper and aluminum. On the other hand, the speed of the machining of titanium is about 10 to 100 [1, 2] times slower than machining of aluminum and in cases where the speed of production is a priority, the use of titanium is not profitable. One of the problems with welding titanium alloys is formation of a frangible structure. Factors such as type of welding process and the degree of cleanliness of the surface of the components and the degree of protection are effective on forming this defect and other defects of titanium welding. Researches have shown that by changing the essential variables, Ti-6Al-4 V is one of the most commonly used alloys of titanium due to its combination of desirable physical and mechanical properties. In order to improve the mechanical properties of this alloy, thermal operation is carried out on it [1, 2]; but during welding these properties change with resizing of grains, the ratio of phases, morphology and microstructure. It should be possible to design processes, by taking into account the appropriate variables that can maintain the desired properties of these alloys and even improve some of them.

In welding of titanium alloys, GTAW or TIG (Gas Tungsten Arc Welding or Tungsten Inert Gas) process is widely used because of the simultaneous provision of high quality at low actual cost. The basis of this method is to create a weld bonding by making electric arc with a non-consumable tungsten electrode and protecting the molten pool through a protective gas stream. In tungsten inert gas welding, the grains are stretched [1, 2]. As a result, mechanical properties such as toughness and ductility are reduced. Due to high activity of this metal, the protection of the weld area is carried out even up to a temperature of 426° C [1, 2]. In the conducted researches, the ultimate tensile strength of the weld area of this alloy has been reported between 925 and 1060 MPa [1, 2]. Also, in comparison to the strength of the weld area and the base material, contradictory results have been reported. That is, both lower strength and greater strength of the weld area than the base material have been reported. This is due to effect of the type of welding process and the thermal operation after welding on the final properties of the weld area. The martensitic microstructure, which is formed in high-energy welding processes, has a higher hardness and strength than the products of this arc welding process. Ductility of weld area is less than the base material. Increase in length of this alloy is 11% in melting mode and 16% in the worked mode. However, this quantity has been reported less than 6% [1, 2] for its weld area. In welding of this alloy the least degree of ductility is achieved in method of tungsten inert gas (TIG) and the most degree of that is achieved by electron beam welding. By performing thermal operation after welding, this low ductility can be increased, though not as much as metal-base ductility. The weld area is strongly affected by the beta phase grain size. So, in some cases, by performing a TIG process using pulse current, the beta grain size can be reduced and thus toughness can be increased. In this alloy, by controlling microstructure the fracture toughness and the resistance to fatigue cracking in better weld area can be usually improved even compared to the base metal, which is achieved through the formation of a fine layer microstructure.

As other welding processes, the properties of the final product of GTAW welding are influenced by variables of this process (such as intensity of welding current, cooling rate, welding rate and protective gas flow, porosity in the weld area also varies). Some of these variables are welding current, voltage, electrode type and diameter, chemical composition and protective gas flow rate, welding rate, and so forth. By changing these variables, the ultimate weld properties can be changed. It should be noted that sum of effects of these variables can be found in the quantitative effectiveness called heat input. In other words, each of these variables can directly or indirectly affect the heat input and thus affect the ultimate property of weld. This quantity represents the amount of energy transferred per weld length unit. Voltage, current and welding rate directly affect this quantity, and parameters such as diameter of electrode by affecting the relative density of the current and the neutral gas flow rate by changing the heat transfer around the weld, indirectly affect this quantity. Excessive heat input can cause problems such as weld cracking and reduction in mechanical properties of the piece.

In this research, after selecting the intended variables and their change levels, welding operation was performed on alloy sheets of Ti-6Al-4 V through GTAW method. In the end, the effect of variables of current intensity and welding voltage on the properties of the final product was investigated by examining the depth of welding penetration, the mechanical properties of the weld, the microstructure and the base metal of the weld area and its surrounding areas. In addition, the cost comparison of manual and automatic welding processes has been put in **Table 1**.

1.1 Purpose of the research

The purpose of this study was to investigate the effect of electric current intensity in the process of arc-welding with tungsten electrode. Due to the fact that titanium metal is widely used in advanced industries such as aerospace and it is also difficult to welding with conventional methods, this material has been used for welding in this study. In this research, the process of welding arc-welding with tungsten electrode (GTAW) was used to weld titanium sheets. Preparation of the welding seam with V shape was carried out, and it was used of the protective gas of the melting pool and the secondary protective gas and back protective gas to provide complete protection

Method of welding	Manual	Automatic
Welding Current (amps)	240	300
Welding Voltage (volts)	25	30
Travel Speed (in/min)	10	15
Gas Flow (f^3/hr)	30	40
Welding Time (hr)	0.057	0.022
Arc Time (hr)	0.02	0.013
Labor + Overhead Cost ($/hr)	18	18
Operator Factor (%)	35	60
Weight of Deposit (lbs)	0.037	0.037
Filler Wire Cost ($/lb)	1.6	1.6
Deposition Efficiency (%)	100	100
Gas Cost ($/$f^3$)	0.06	0.06
Gas Used (f^3)	0.6	0.53
Electric Power Cost ($/kw-hr)	0.035	0.035
Power Source Efficiency (%)	50	50
Labor + Overhead Cost ($/ft)	1.026	0.396
Filler Wire Cost ($/ft)	0.059	0.059
Shielding Gas Cost ($/ft)	0.036	0.032
Electrode Cost ($/ft)	0.001	0.001
Electric Power Cost ($/ft)	0.008	0.008
Total Cost ($/ft)	**1.130**	**0.496**

Table 1.
Cost comparison of manual vs. automatic welding [3].

during the welding. At first, the welding process was performed with three electric current of 80, 90 and 100 amps. Then, PQP samples were prepared according to ASMEM sec IX standard, and tension, impact and hardness tests were performed on them.

2. Review on references

2.1 Titanium and its alloys

The pure titanium has a high coefficient of friction and a low hardness; therefore, it has low wear resistance. The density of titanium is about half of steel. Although titanium has the highest strength-to-weight ratio among highly used industrial metals, but its high price limits its use. Titanium has a strong tendency to react with oxygen. High corrosion resistance of this alloy is due to formation of a sticky oxide layer at its surface. Incompleteness of the last titanium electron layer has caused the element to form solid solvent by creating ionic and covalent bonds with many elements. Titanium alloys are used in the aerospace, chemical and medical industries due to their corrosion resistance, high strength-to-weight ratio and maintenance of strength at high temperatures [1, 2]. **Table 2** shows the important properties of titanium with other metals widely used in the industry.

Titanium is transformed from phase α which has HCP structure to phase β with BCC structure at 882° C. The phase transformation makes the possibility of creating different thermal operations for obtaining various microstructures. Schematic description of titanium phases and their main transformation temperatures are shown in **Figure 1**.

The classification of titanium alloy elements is based on their effect on the phase transformation of α to β. The temperature of this phase transformation depends on the type and amount of elements in titanium alloys. From this perspective, these elements are divided into three groups of neutral, phase stabilizer α and phase stabilizer β. This division is presented in **Table 3**.

	Ti	Fe	Ni	Al	Cu
Melting point (C °)	1670	1538	1455	660	1084
Phase transformation	α to β	α to γ	—	—	α to β
Phase transformation temperature (C °)	882	912	—	—	810
Crystal structure	BCC، HCP	FCC، BCC	FCC	FCC	FCC
Elastic Modulus (GPa)	115	215	200	72	125
Degree of yield strength (MPa)	1000	1000	1000	500	1000
Density (g / cm^3)	4.5	7.9	8.9	2.7	8.94
Corrosion resistance	Very high	Low	Medium	High	High
Reactivity with oxygen	Very high	Low	Low	High	Low
Relative price	Very high	Low	High	Medium	Medium

Table 2.
Comparing properties of titanium with other important industrial metals [4].

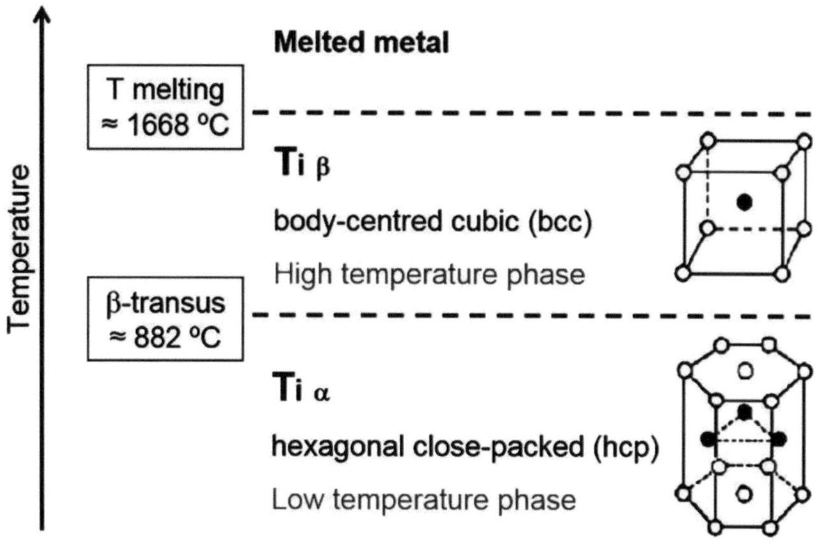

Figure 1.
Schematic description of titanium phases and main transformation temperatures [2].

Type of element	Name of element
Neutral	Tin, zirconium
phase stabilizer α	Aluminum, carbon, nitrogen, oxygen, germanium
β Isomorph	Molybdenum, Vanadium, Tantalum, Niobium
β Eutectoid	Iron, manganese, chromium, cobalt, nickel, copper, silicon, hydrogen

Table 3.
Classification of titanium alloy elements [1].

Neutral elements do not have much effect on this transformation; while phase-stabilizing elements of α stabilize phase α to higher temperatures than transformation temperature, and phase-stabilizing elements of phase β cause the phase to be stable at lower temperatures than transformation. Stabilizers of phase α, in addition to stabilizing this phase, also expand two-phase area of α + β at higher temperatures than the transformation temperature. Stabilizers of phase β are divided into two categories. The first group creates a two-part isomorphic system with a relatively good solubility. The other group creates a eutectoid point in the phase diagram and forms intermetallic compounds [1, 5, 6]. Phase diagrams of titanium based on the type of alloying elements are shown in **Figure 2**.

Titanium alloys, in terms of microstructure, are divided into three main groups of α, β and α + β, and two subgroups of close to α and close to β (pseudo-β). Alloys containing stabilizing elements of phase α and neutral elements are included in α alloys. By adding small amounts of β-phase stabilizers to α alloys, α alloys are formed, and subsequently the phase mixture of α + β alloy is formed, by increasing the stabilizing elements of β, if the amount of these elements is sufficiently large, phase β is formed lonely, and if it is not enough to make martensitic phase by quenching the alloy, it is called pseudo-β alloy. α alloys have low relative strength compared to β, but their properties are well maintained at high temperatures. Shaping of two-phase α + β

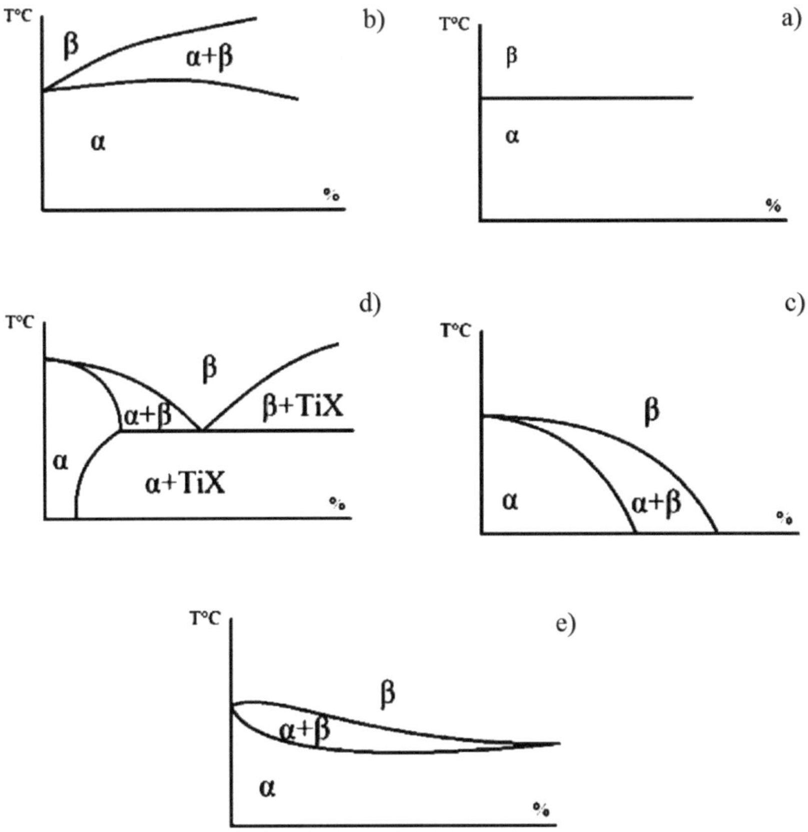

Figure 2.
Phase diagrams of titanium based on the type of alloying elements; a) pure titanium, b) stabilizer α, c) isomorph β, d) β-eutectic, e) neutral [7].

alloys is more convenient than α alloys, but it is more difficult than β alloys [6–8]. **Table 4** shows the important commercial titanium alloys of α and α + β, and **Figure 3** shows the phase stability diagram of titanium alloys based on concentration of stabilizing alloying elements of phase β.

2.2 Bonding of titanium and its alloys

One of the most important methods for bonding these alloys is fusion welding, non-fusion welding, such as friction welding and hard soldering. Due to very high activity of titanium, its bonding processes should be carried out under specific conditions, so that the molten pool and also solid titanium are protected well. For fusion welding of titanium alloys, five main methods are used including arc welding with tungsten electrode (non-consumable electrode), arc welding with metal electrode (consumable electrode), electron beam welding, plasma arc welding and laser welding; among these methods, TIG method is more common [9–14]. Of course, in each of these melting methods, there is the risk of contamination of the alloy by elements such as oxygen, nitrogen, and hydrogen. On the other hand, the production of suitable fillers for titanium alloy titanium welding has problems. First, it is very hard to produce wire of titanium alloys, which has both high strength and relatively low ductility. Secondly, in the case of producing these wires, the actual cost is too

Common name	Alloy composition (wt %)	$_{\beta}T$ (C)°
	α alloys and commercial pure titanium	
Grade 1	CP-Ti (0.2Fe, 0.18O)	890
Grade 2	CP-Ti (0.3Fe, 0.25O)	915
Grade 3	CP-Ti (0.3Fe, 0.35O)	920
Grade 4	CP-Ti (0.5Fe, 0.40O)	950
Grade 7	Ti-0.2Pd	915
Grade 12	Ti-0.3Mo-0.8Ni	880
Ti-5-2.5	Ti-5Al-2.5Sn	1040
Ti-3-2.5	Ti-3Al-2.5 V	935
	α + β alloys	
Ti-811	Ti-8Al-1 V-1Mo	1040
IMI 685	Ti-6Al-5Zr-0.5Mo-0.25Si	1020
IMI 834	Ti-5.8Al-4Sn-3.5Zr-0.5Mo-0.7Nb-0.35Si-0.06C	1045
Ti-6242	Ti-6Al-2Sn-4Zr-2Mo-0.1Si	995
Ti-6-4	Ti-6Al-4 V (0.20O)	995
Ti-6-4 ELI	Ti-6Al-4 V (0.13O)	975
Ti-662	Ti-6Al-6 V-2Sn	945
IMI 550	Ti-4Al-2Sn-4Mo-0.5Si	975

Table 4.
Important commercial titanium alloys of α and α + β [5].

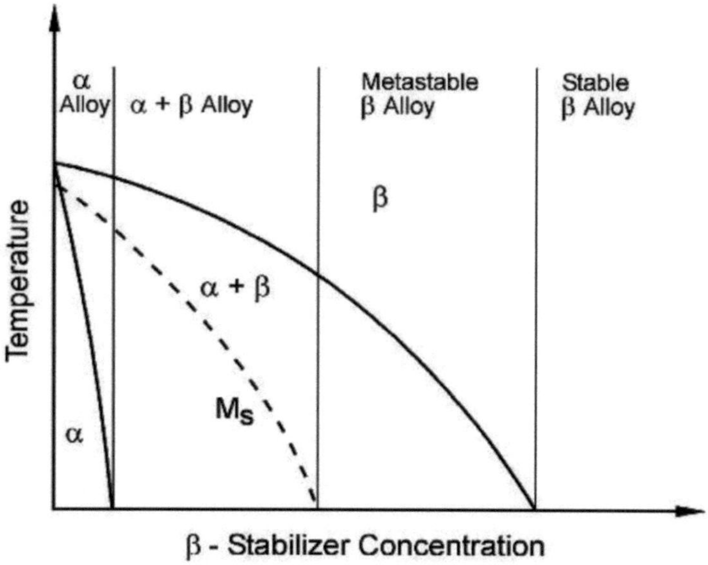

Figure 3.
Phase stability diagram of titanium alloys based on concentration of stabilizing alloying elements of phase β [5].

high. MIG (consumable electrode) method is used when the welding seam is large and the filler wire is selected homogeneous with the consumable electrode. Plasma arc welding was proposed as an alternative to TIG method. The area affected by heat in this method is narrower than the same area of the TIG method. In this method, the thermal source for metal melting is plasma torch [14–17]. The specifications of the four methods for fusion welding of titanium alloys are shown in **Figure 4**. According to the figure, in methods based on high-energy beams such as laser method (LBW) and electron beam method (EBW), the weld area is deeper and narrower than the arc method. In addition, their heat input is less than arc method. This less heat is equal to residual stresses and less distortion in the workpiece.

For successful welding of titanium alloys, it should be noted that titanium is highly active and reactive, especially at temperatures above 500. This material can react with elements in impurities or air such as carbon, oxygen, nitrogen, and hydrogen. Although these elements can increase titanium strength under certain conditions, but they can reduce the ductility and tensile strength of titanium connections. On the other hand, along with high reactivity of titanium, effects of heating and cooling cycles that affect mechanical properties and chemical composition of titanium connections during welding processes also need to be considered [8, 19].

TIG process can be done either manually or automatically, and the heat source and welded wire can be independently controlled. The process can be done depending on the conditions, either using a welded wire or without it. The connections created by TIG method usually have high quality and the amount of defects, splashing (spatter) and slag is low, which is why cleaning after welding is usually not considered a key step. One of the other advantages of this method is the relatively inexpensive equipment of that. On the other hand, the welding rate in this method and the sedimentation rate are low, which is why this method is not efficient for welding thick parts. This process is sensitive to the contamination in the weld wire and the base metal, and the magnetic field causes deviation of the arc. All of these factors make controlling the process difficult. Also, the heat input is very high [20–22].

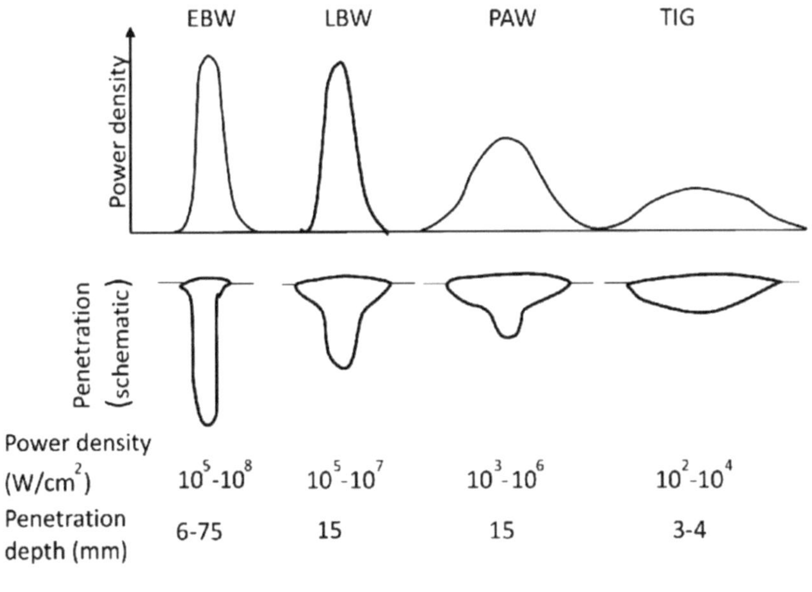

Figure 4.
Comparing specifications of welded titanium specimens by fusion methods [18].

2.3 Problems with welding of titanium alloys

Welding of titanium and its alloys has some problems. Some of these problems include:

- Preventing the mixing of the molten at the weld edge due to formation of oxides,
- Cracks,
- Creation of porosity
- Presence of oxides in the form of offal in the weld metal,
- Although dissolution of oxygen and nitrogen in titanium increases its hardness, but dissolution of hydrogen causes loss of toughness and increase of sensitivity to crack of the alloy.
- Formation of highly stable carbides in the event of reaction wit carbon at high temperatures; although in some cases, small amounts of carbon are added to this alloy to obtain the alloy with high hardness and strength, but in general it can be said that formation of these carbide phases makes the alloy frangibility [23–25].

2.4 Ti-6Al-4 V

Toughness of the two-phase α + β alloys is high. An optimal combination of toughness and fatigue strength in these alloys is created by thermo mechanical operation. The most important alloy of this group is Ti-6Al-4 V or IMI 318. The alloy, used in the manufacture of jet engine fan blades, accounts for more than half of sales of titanium alloys, and the alloy is used in more than 80 percent of applications related to aerospace of titanium alloys. In cases where high corrosion resistance is considered along with high strength and toughness, this alloy is used. In this alloy, aluminum and vanadium are respectively, stabilizers of phase α and β. In the industry, this alloy exists as two grades of 23 or ELI and Grade 5. The chemical composition of this alloy is shown in **Table 5** and its physical and mechanical properties are given in **Table 6**.

Due to formation of hard and needle-like phase of martensite and large grain size of phase β in the penetrative area, which reduces its ductility, this alloy does not have much suitable capability of weldability. Generally, to improve the properties of the weld area, the grain size of phase β is tried to be decreased and the amount of heat input is also controlled at an optimum size. Comparison of the tensile strength of this material at various temperatures with other materials is shown in **Figure 5**. As it can be seen, in the range of working temperature of 400–500° C, this alloy has the highest specific strength among different materials [11, 25, 27, 28].

Name of element	Al	V	Fe	O	N	C	H	Other impurities	Total other impurities
Min wt%	5.5	5.3	—	—	—	—	—	—	—
Max wt%	6.75	4.5	0.3	0.2	0.03	0.08	0.125	0.1	0.4

Table 5.
Chemical composition of Ti-6Al-4 V [8].

Property	Value
Density	g/cm^3 4.43
Melting point (°C)	1604–1660
Elastic modulus	GPa1113.8
Thermal conductivity	W/m.K 6.7
Electrical resistance	0.000178 Ohm-centimeters
Thermal expansion	/K8.9*10–6
Tensile strength	MPa 880
Compressive strength	MPa 970
Vickers Hardness	334

Table 6.
Physical and mechanical properties of Ti-6Al-4 V [26].

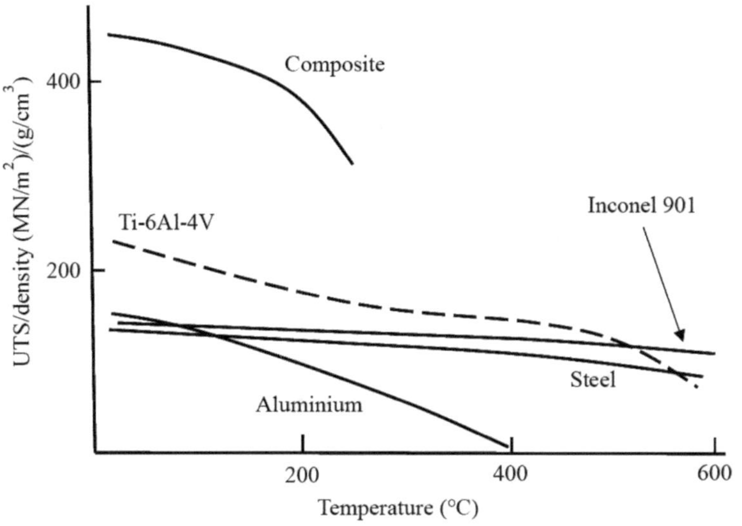

Figure 5.
Comparison of the tensile strength of Ti-6Al-4 V with other materials at various temperatures [18].

2.5 Manual electrode arc welding and protective gas

2.5.1 History of tungsten arc welding

After the discovery of short arc in 1800 and continuous electric arc in 1802, electric arc welding was slowly developed. The idea of welding in inert gas was introduced in 1890. But even in the early 20th century, welding of non-ferrous materials, such as aluminum and magnesium, was still difficult. Because these metals react quickly with air and thus are oxidized and strength of the metal is reduced. The process of using flux as an electrode coating was also not satisfactory and did not protect against contamination. To solve this problem, in 1930, inert gases were used. A few years later, welding under protective gas was used for the first time in aviation industry for welding of magnesium, and in this operation the electric arc formation method was utilized using helium and argon as protective gases and favorable results were obtained [21, 29].

2.5.2 Arc welding with tungsten electrode and inert gas

It is a type of fusion welding which is classified according to DIN 1910 into two general types of TIG welding and arc welding with protective gas or GMAW. The equipment used in the TIG procedure is much cheaper than other methods. This method can be used to weld pieces with thicknesses of 2 to 20 mm with proper efficiency. This method usually uses a current between 50 and 500 amps and a voltage between 10 and 15 volts. The type of electrode is determined according to the desired properties and its diameter is determined proportional to the thickness of the piece and the welding condition. The schematic of TIG welding process is shown in **Figure 6**.

In TIG welding, also known as GTAW, non-consumable tungsten electrode is used which is protected by gas or a mixture of gases. This electrode can be pure tungsten. To increase the arc stability and make it easier to start the arc, a tungsten electrode containing small amounts of oxides such as zirconium oxide, thorium oxide, etc. can be used [13, 14, 19, 30–32]. Specifications of tungsten electrodes are given in **Table 7**.

The sedimentation rate in TIG method is low. This sediment rate can be improved using preheated filler. Due to the fact that the electrode in TIG method is non-consumable, therefore, depending on the thickness of the workpiece, type of connection and ... filler metal also can be used for welding (filler or weld wire). These fillers can be added manually or mechanically to the weld area [18, 33]. Application fillers for TIG welding of titanium and its alloys are listed in **Table 8**. This classification is made by the American Welding Society (AWS).

The created arc contains a plasma environment that provides the required energy to melt the base metal and filler. This plasma environment is created through an electrical discharge between the electrode and the work piece. To create such a good discharge, a high-frequency and about few MHz ignition system is used. The temperature of the arc above the boiling pond area is between 1200 and 1500 Kelvin and at the molten surface it is between 1700 and 2500 Kelvin [35–37].

Although GTAW welding method has advantages over other methods, but compared to other welding methods, the following disadvantages can be noted:

- Lower welding rate than some methods,
- Lower metallurgical quality of the resulted weld than some methods,
- High heat input, which leads to large distortion and the area under the influence of the heat gets bigger,
- Sensitivity to changes in arc length [9–11].

2.5.3 Polarity of welding

In this method, electrical energy is transmitted from the welding torch to the base metal by electric arc. This kind of welding process can be divided into three categories in terms of the used current:

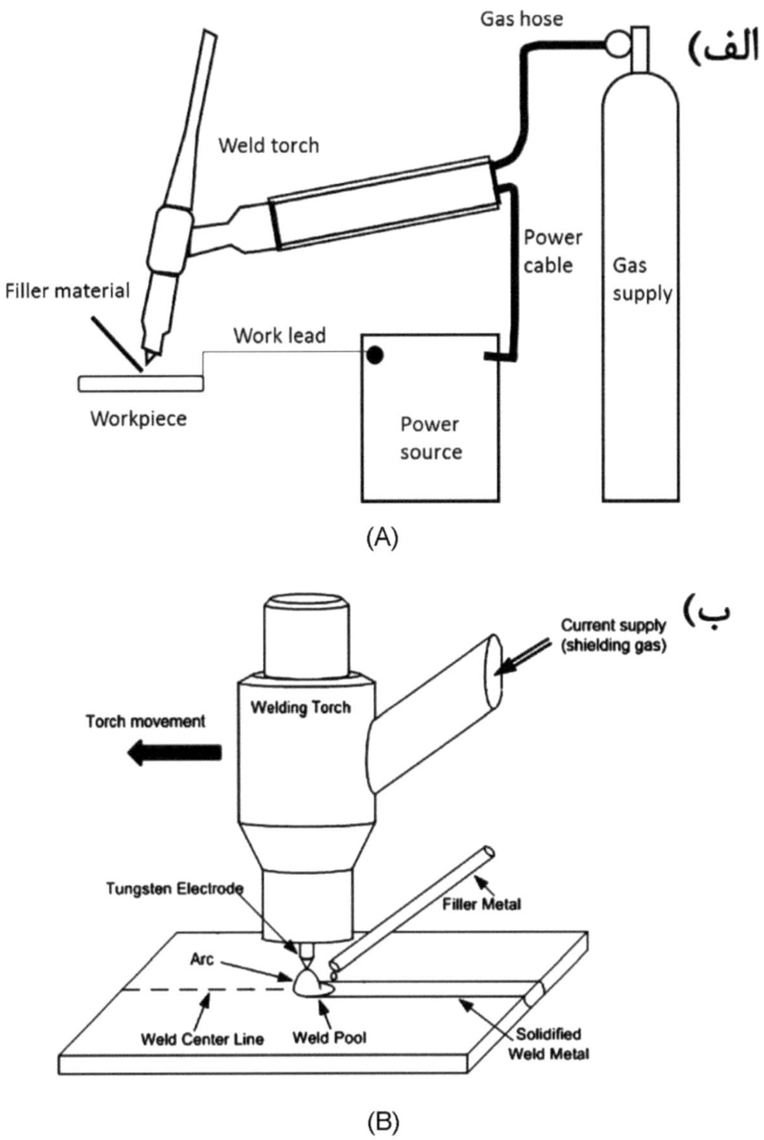

Figure 6.
Process of GTAW; A) schematic of the system used in the process, B) welding principles [7, 18].

- DC Current - Negative-Pole Electrode: This is also called direct polarity mode and is very common in GTAW process. The electrode is connected to the negative pole of the power source. The electrons are emitted from the tungsten electrodes and accelerated passing through the electric arc.

- DC Current - Positive-Pole Electrode: This mode is called Reverse Polarity. The electrode is connected to the positive pole of the power supply. In this situation, the thermal effect of electrons in tungsten electrode is greater than the work piece. Because electron bombardment is toward the electrode. Therefore, an electrode with a large diameter and an anhydrous system is required to be prevented the melting of the tip of the electrode. Positive ions bombard the surface of the work piece, which causes the oxide shells to break down on the work piece. Therefore, this can be used to weld materials with hardened oxide layers such as aluminum and magnesium.

- AC Current: In this case, the penetration and cleaning of the work surfaces of the oxide layers are done well. This is often used for aluminum welding.

Figure 7 shows the effect of polarization systems used in TIG welding processes on the shape of the weld area.

Two other types of currents are pulsed and non-pulsed currents. If there is a need for a very high penetration, pulsed current is used. Direct polarity is the most commonly used polarity in GTAW processes. Through this polarity, high heat in the work piece and, as a result, a high penetration and narrow weld area can be achieved. By choosing alternative current, a combination of cleaning of oxidation from the surface and the proper penetration can be achieved. In TIG welding of titanium alloys, arcs with direct current and tungsten non-consumable electrode as negative pole are

Grouping	Color	Alloying element	Alloy oxide	Nominal weight percent of alloy oxide
EWP	Green	—	—	—
EWCe-2	Orange	Serum	CeO2	2
EWLa-1	Black	Lanthanum	La2O3	1
EWLa1.5	Golden	Lanthanum	La2O3	1.5
EWLa-2	Blue	Lanthanum	La2O3	2
EWTh-1	Yellow	Thorium	ThO2	1
EWTh-2	Red	Thorium	ThO2	2
EWZr-1	Brown	Zirconium	ZrO2	0.25

Table 7.
Tungsten electrodes and tungsten alloys [33].

Weight percent element	ERTi-1	ERTi-2	ERTi-3	ERTi-5	ERTi-7
Nitrogen	0.015	0.020	0.020	0.030	0.020
Carbon	0.03	0.03	0.03	0.05	0.03
Hydrogen	0.005	0.008	0.008	0.015	0.008
Iron	0.10	0.20	0.20	0.22	0.20
Oxygen	0.10	0.10	0.10–0.15	0.12–0.20	0.10
Palladium	—	—	—	—	0.12–0.25
Aluminum	—	—	—	5.5–6.75	—
Vanadium	—	—	—	3.5–4.5	—
Titanium	Rest	Rest	Rest	Rest	Rest
Base material of ASTM	Gr. 1	Gr. 2	Gr. 3	Gr. 5	Gr. 7

Table 8.
Fillers used in welding of titanium alloys [34].

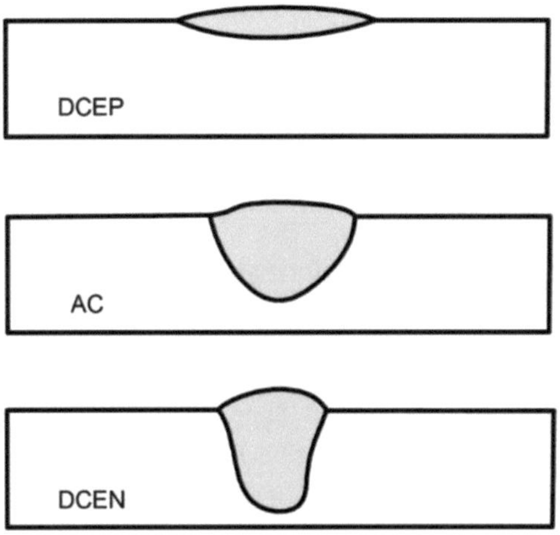

Figure 7.
Effect of current type and polarity on the shape of the weld area [38].

usually used. It should be noted that if there is surplus current, tungsten electrode welding is melted and the tungsten offal will remain in the weld area [10, 14, 21].

2.5.4 Inert gas

The purpose of using inert gas is to create a stable arch, to protect the molten pool and tungsten electrodes against environmental contamination. Some of these gases are argon, helium and their mixture. In welding of major materials, argon is used. Since adding helium to argon improves heat transfer, in welding thick pieces the mixture of these two gases are used. Since argon has low ionization voltage and it is ionized easily, provides this possibility for the arc to be easily arranged and remain stable; therefore it is suitable for working with AC current. Also, argon makes start of arc in AC current easier. Argon generates a column of concentrated arc, and has less capability of thermal conductivity than other gases. Since argon causes stabilization (keeping the arc fixed), it is used in most of mixture of protective gases. The mixture of argon and helium is also used in TIG process when both high penetration and slow arc is considered. The density of argon is approximately same with air, so the same amount of gas that enters the system makes the coating, but the density of helium is one tenth of the air. Therefore, argon should be entered the system 4–5 times more to create the same amount of protective coating. The first ionization energy of helium is extremely high, so the temperature needs to be raised to a large extent. This causes the created arc to be very hot compared to argon. The thermal conductivity of helium is much more than argon; therefore, using helium, the created heat in the center of the arc moves easily and creates a bowl-like area (more open pool). Helium is typically used in industrial processes and argon is used in TIG manual welding. Argon is also much cheaper than helium [5, 21, 39, 40]. Effect of the protective gas composition on the weld geometry and properties of protective gases used in TIG welding are respectively shown in **Figure 8** and **Table 9**.

2.5.5 Heat input

Heat input is a relative quantity that expresses the amount of energy transmitted per length unit of weld. This quantity is directly affected by the mechanical properties

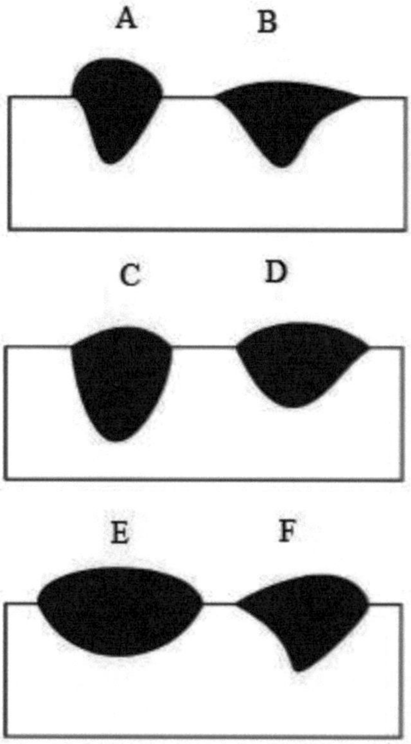

Figure 8.
Effect of the protective gas composition on the geometry of the weld; A) argon; B) mixture of argon and oxygen; C) carbon dioxide, D) mixture of argon and carbon dioxide, E) helium, F) mixture of argon and helium [38].

Gas	Chemical Symbol	Molecular Weight (g/mol)	Specific Gravity of Air	Density (g/L)	Ionization Potential (eV)
Argon	Ar	39.95	1.38	1.784	15.7
carbon dioxide	CO_2	44.01	1.53	1.978	14.1
Helium	H_e	4.00	0.1368	0.178	24.5
Hydrogen	H_2	2.016	0.0695	0.090	13.5
Nitrogen	N_2	28.01	0.967	1.25	14.5
Oxygen	O_2	32.00	1.105	1.43	13.2

Table 9.
Properties of protective gases used in TIG welding [23, 35].

and metallurgical structure of the weld area and the area affected by heat. The following equation is used to calculate the input heat:

$$H = \frac{60\ \mathrm{EI}}{1000\ \mathrm{S}}$$

where its variables include H: heat input based on Kj/in or kj/mm, E: arc voltage based on volt, I: current based on ampere and S: speed of movement based on in/min or mm/min.

TIG method has a large amount of heat input during welding, and the product of this welding has a large grain size. However, methods such as laser and electron beam welding have fewer heat input than TIG method, and thus reduce the size of the β-phase grain. But these methods have high cooling rates after welding, which increase the amount of needle-shaped martensitic phase. The cooling rate of weld is dependent on the input heat. In Ti-6Al-4 V, phases and various microstructures can be created under different cooling rates, these microstructures determine the final properties of the weld. Therefore, when using these methods, after performing the welding, appropriate heat operation is needed to enhance the ductility of the penetrative area of weld and the area affected by heat. Of course, reducing the amount of input heat is desired to the extent that it does not interrupt the penetration value [35, 41, 42].

2.6 Welding and mechanical properties of Ti-6Al-4 V

The most common method of welding this alloy is TIG, and is usually used for welding the narrow piece of this alloy. For entire-layer microstructures of titanium alloys, beta size parameters, size of alpha colony, width of alpha plates and the presence of grain boundary alpha layer are key microstructural parameters. Among these parameters, the size of the alpha colony, which determines the effective length of slip, is more important. By decreasing the size of alpha colony, yield strength, ductility, microscopic crack germination, and crack propagation resistance increase, while macroscopic crack propagation and fracture toughness increase with increasing size of alpha colony. If martensitic microstructure is formed and a slip length equal to the width of an alpha plate is created, the yield strength is significantly increased [4].

The tensile ductility of entire-layer structure is less than the dual structure. Microcavities can germinate in the alpha-beta intersection due to high dimension ratio of the alpha plates. The number of these intersections is high in entire-layer microstructures.

Reducing the initial beta size has a positive effect on tensile ductility. If the difference between strength among grain boundary alpha layer and transgranular microstructure is high, cracks can germinate in grain boundary alpha layer and lead to intergranular fracture [4, 11].

For a welded area, a fine layer microstructure formed in the melting area and the area affected by heat, has higher hardness than the base metal. The final tensile strength of the titanium alloy weld area has been reported between 925 and 1060 MPa. Based on the carried out researches, the weld area can have a greater or lesser strength than the base metal, depending on the type of welding process and thermal operation treatment welding. Of course, the martensitic microstructure formed in processes with a high energy intensity has higher hardness and strength than the weld created by arc welding methods [43, 44]. Ductility of the weld area can be improved by thermal operation. However, its amount cannot be increased as much as the base metal. Ductility can also be improved by reducing the initial beta grain size. One way to reduce the initial beta grain size is to use pulse current in TIG welding, which can lead to increase in ductility [42–46]. In the weld area, due to formation of fine layer microstructure, fracture toughness and growth resistance of fatigue cracks are greater than the base metal [47].

2.7 Welding parameters affecting weld properties

A part of the weld defects can be eliminated by selecting the right amount of welding variables. In addition, given that the input heat amount is strongly effective on metallurgical properties of the weld, and considering the close direct or indirect bonding of variables of welding and heat input, these are in fact the welding variables that affect the metallurgical properties of the weld.. These properties determine the mechanical properties, corrosion and etc. of a piece. As a result, changing the variables can change the mechanical properties of the weld [25, 48, 49].

As already mentioned, the effect of changing the parameters of the welding process should be sought in how they affect the heat input quantity. Changing the diameter of the electrode changes the density of current, melting rate, sedimentation rate and penetration depth. The lower the diameter of the electrode, the higher the current density and as a result the higher the heat input. On the other hand, in terms of the operation and apparent quality of weld, electrodes with higher diameter, by reducing arc instability, reduce the amount of discharge and result in a cleaner work piece [15]. The change in voltage is strongly effective on the width of the weld area. The voltage also affects the heat input. The higher the welding voltage, the higher the heat input [50]. It is predicted that as the current intensity increases, the penetration depth of the weld will increase due to increase in force on the molten drops. It should be kept in mind that the diameter of the electrode should also be considered in choosing the intensity of the welding current. Because as the electrode diameter is higher, the current intensity must be increased to maintain the current density [51]. In various studies, various inert gases have been used to protect the molten area of GTAW welding of steels. In some researches, effect of mixture of inert gases has been investigated. Inert gases, in addition to the main task which is protection, also affect the heat and heat transfer in the protection zone. By changing the chemical composition of inert gases, the protection level varies. Meanwhile, due to the fact that gases have different heat transfer coefficients, by changing the composition, the amount of heat transferred by the protective gas changes. On the other hand, the change in the rate of the protective gas by affecting the amount of heat transferred by the gas is effective on the amount of heat accumulated in the weld area. It should not be forgotten that, since protective gas is one of the bases of GTAW welding method, any change in the protective gas composition or its current rate must be such that it does not interrupt the protection [52, 53].

2.8 Review on conducted researches

In 2003, Joe et al. evaluated the impact toughness of bonding of Ti-6A1-4 V welded through GTAW. They prepared Charpy test specimens in such a way that impact toughness of three areas of base metal, the weld area and the area affected by heat could be compared with each other. Their research showed that microhardness of the weld area was higher than the area affected by heat and impact toughness of this area was also 50% higher than the two other areas. They attributed the improvement of impact toughness to reduction in the initial alpha grains in the weld area. Grain boundaries of initial alpha are considered the preferred placed for germination of microcracks and provide easy routs for crack propagations [54].

In a research in 2008, Balasubramanian et al., with the aim of predicting the tensile properties of welded Ti-6Al-4 V, applied mathematical modeling of the pulsed current technique. This modeling was done by taking into account 4 variables in 5 levels of change and designing the test by response surface methodology. The efficiency of the model was verified using analysis of variance (ANOVA table). They provided an efficient model with a confidence level of 99% to predict the obtained weld tensile properties [31].

In another research in 2011, Balasubramanian et al. evaluated the growth of fatigue cracks of Ti-6Al-4 V welded through TIG method, electron beam and laser beam. They observed the highest and the lowest growth resistance of fatigue crack in the made weld, respectively, through laser beam method and TIG method. In addition, the growth resistance of fatigue crack was reduced after welding on this alloy. They attributed increase in growth resistance of the made weld though laser beam method to fine layer microstructure in the weld metal caused by lower input heat and higher cooling rate [32].

In the same year, Balasubramanian et al. examined the effect of type of welding method on the microstructure and mechanical properties of Ti-6Al-4 V welded through TIG, electron beam and laser beam. The weld strength made by electron beam welding was 6% and 2% higher than TIG and the laser beam. This is while the toughness of the weld made through TIG method was 35% higher than the other two methods due to presence of needle-shaped phases in grain boundaries and Widgetman Stroten's vaporization of $\alpha + \beta$-biphasic area. They also reported the Vickers' hardness for TIG and electron beam welding, respectively, 403 and 509, and stated that by correcting the microstructure of the weld metal, it is possible to make changes in mechanical properties [55].

In 2011, Chen and Pan conducted a research using dynamic controlled plasma arc to minimize the heat input to a workpiece of Ti-6Al-4 V while maintaining full penetration.

They reported increase in toughness, hardness and ductility in the weld area due to reduction in beta-phase grain size, reduction in heat input and inhibition of making the hard martensitic phase in the penetrative area [9].

In 2015, Bohorquez and Cunha carried out a review study on ultrasonic methods in arc welding of alloys of titanium, steel and aluminum. They conclude improvement in the performance of arc welding processes at industrial scale using pulsed arc welding using the ultrasonic method, especially for alloys such as titanium alloys which have weldability problems [12].

In 2017, Singh et al. conducted a review study on methods for improving the penetration depth of weld in TIG welding of commercial pure titanium, aluminum and stainless steel. They focused on finding ways to eliminate defects in the low penetration depth of TIG welding and improving the mechanical properties of the resulted welds, and they knew use of flux (slag) or pulsed current effective to improve the penetration depth [13].

In 2017, Yang et al. welded the pulsed plasma arc of Ti-6Al-4 V. They reported a 24 to 30 percent decrease in grain size in the penetrative area, and 68 percent increase in the elongation quantity at frequency of 20 kHz, and a 38 percent increase in the elongation quantity at frequency of 40 kHz. In addition, in the end they introduced frequency range of 20–40 kHz as the best frequency range for the welding of this alloy [14].

2.9 Methods for evaluating properties of weld area

2.9.1 Metallography

In metallurgy, it is very important to observe, evaluate and determine the microstructure of phases or the components of the material, since it is possible to distinguish many properties of material and justify many of its behaviors under various conditions by microstructures. To achieve this, a metallographic method is used. Some uses of metallography is the awareness of the chemical composition and different properties of materials, detection of crystalline building, the history of mechanical work and the history of thermal operation carried out on the sample. To do a metallography, sample must be prepared first. The first step is to cut the piece that we want to prepare the sample of that. We use a cutter apparatus to cut the piece. This apparatus, while cutting, creates a coating of water on the piece because the collision of the cutter with the piece creates friction and increases the temperature of the cutting location during cutting, and if this temperature does not drop, the phase structure around the cutting location may change. To solve this problem, water is used up to decrease the temperature. Till now, the sample has been separated from the piece. In the next step, mounting muse be done. After preparing the mount, it is time for sanding; the sandpaper itself consists of 4 rows with numbers 240, 360, 600 and 1000. The sandpaper is soaked by water. The concept of 240 is that it has 240 holes (groove) per inch in square. First we use sandpaper 240. We select a specific direction on the sample and then sand toward that direction. By changing the sandpaper, this time, we will sand the sample in a perpendicular direction to the previous state. When all the lines are in one direction, we go to the next sandpaper. After sanding, we will thoroughly wash the sample with water. The next step is polishing, which uses a suspension (0.3 micron) of Al_2O_3. We pour it on a rotating disk fabric. To polish, one of the following three methods is used: we rotate the sample on the inverse rotation direction of the disk, or rotate it as shape of 8, or rotate it from side to middle of the disk, while moving it linearly, simultaneously. We do this 20 times each for 30 seconds. Then we wash the sample with water thoroughly. In the third step, we use 0.05 micron Al_2O_3 and a softer fabric on the disk. The pressure of the hand should be less than the previous step. When the sample lines disappear, we wash it with water and then alcohol. Then we dry it with a tissue paper and a heater. Now, we see the sample under a microscope. If there are many lines, we repeat one or more stages of the preparation process, depending on the number and depth of the lines, because these lines block the correct view of the sample, and sometimes lead to mistakes in conclusion and report of observations. The prepared sample does not show a particular image of the microstructure. To see the microstructure, we need to etch the sample (chemical engraving). Then we see the structure under the microscope [56–58].

2.9.2 Hardness

We consider hardness as a material resistance to plastic deformation. Although it is conceptually different to strength, but they both operate in the same direction. That is, if the hardness increases, strength increases and vice versa. The strength is a specific number but the hardness is relative. In the sense that the hardness of a material is measured relative to another. There are various definitions of hardness that vary depending on the application. These definitions are as follows:

- Resistance to penetration under static or dynamic forces
- Energy absorption under impact forces
- Scratch resistance
- Abrasion resistance
- Resistance to cutting of machining or drilling

There are different methods to get the hardness value. But in general, there are three general scale types of hardness: scratch hardness, hardness of return or reflection, penetration hardness. The technique of scratching is mostly used in mineralogy. The intended object is scratched with different materials. If the scratch does not occur, it shows that the object of is harder than the material used to scratch, and if not the opposite. This method is very old. Dynamic hardness test is a non-destructive method. A ball bearing is located inside a cylinder whose surfaces are burnished. The initial height of the ball bearing is specific. The cylinder is put on the surface that they want to test the hardness. The ball bearing falls and the hardness value of the material is measured based on the return height. The softer the matter, the more energy it absorbs, and therefore the lower the return height. The hardness in this case is as the unit of energy. Regarding the use of the third type of hardness test method, the types of hardness test are expressed against the penetrating object in following. In the hardness test method against the penetrating object, material resistance to a ball bearing or pyramid with specified dimensions and determined force is considered as hardness. The higher the resistance, the higher the handedness. This test is performed in different ways, which are categorized according to the type of penetrator and the applied pressure:

- Brinell

n this method, a steel ball bearing with 10 mm of diameter, 3000 kg of force for hard metals and up to 500 kg for soft metals are used. Also a tungsten carbide (WC) ball is used for hard metals. Usually, time of applying the force is between 20 and 30 seconds for iron alloys and 65 seconds for non-ferrous metals. The Brinell hardness number is expressed according to load on the internal area of the penetration. The advantage of using Brinell hardness test is that it is less susceptible to surface roughness and scratches than other hardness tests, and also the relatively large size of Brinell effect is useful in obtaining average local inhomogeneities. On the other hand, this large size of the effect can make it impossible to use this test for small pieces or pieces that are under critical stress and formation of penetration can cause fracture in them. Damage in this method is more than other methods.

- Meyer

In this method, instead of the area of the surface within the penetration, the effect surface is used. The average pressure between the penetrating surface and penetration is equal to the force divided by the effect surface. Meyer considered this mean pressure as the hardness, which is a more reasonable value than Meyer hardness.

- Vickers

In the Vickers hardness test, a square-base pyramid is used. The angle between the dimensions is 136 degrees. The reason for choosing this angle is that it is the best ratio of penetration diameter to the ball diameter in Brinell test. Due to penetrating shape, this test also known as the Diamond-Pyramid hardness test. The Diamond Pyramid Hardness Number (DPH) or the Vickers Hardness Number (VH or VHN) is obtained by dividing the force into the penetration area from the following equations:

$$VHN = \frac{2PSin\left(\frac{\theta}{2}\right)}{L^2} = \frac{1.854P}{L^2}$$

In the equation, L stands for average of diameters and P stands for the applied force. Vickers hardness test is acceptable for research works. Because it conducts a continuous test of hardness with a specific load from very soft metals with DPH equal to 5 to very hard materials with DPH equal to 1500. In this mode, the force varies from 1 to 1200 kilograms. Speed of this method is lower and is mostly used for hard materials. Error is also high in this method because error is high in measuring diameter of the effect. On the other hand, since in this hardness test the applied force does not depend on hardness of material so it is better than Brinell and Rockwell.

- Rockwell

In this hardness test method, a diamond-cone mandrel with a vertex angle of 120 degrees or a steel ball bearing of 1.16 or 1.8 inches is used. Since in this method the measurement error is low and speed of the action is high and the created effect is small, so it has many usages. At the beginning, a force equal to 10 kilograms is applied, which makes the surface need less readiness, and the tendency to create penetration or protuberance by mandrel. The application of this force also results in the loss of the oxide layer of the surface, which leads to better result, and then the main load will be applied. The hardness value in this case is given by the apparatus itself, and it is not necessary to measure the diameter of the effect and find the hardness number from the table [39, 59, 60]. In **Table 10**, the specifications of Rockwell Hardness Tests are presented.

2.9.3 Tensile test

One of the most important properties of material is resistance to normal tensile stress. When a piece is affected by tensile stress, it does undergo elastic deformation

Type of Rockwell	Type of Penetrator	Force (Kg)	Usage
A	Diamond-Cone	60	Cemented carbides. Steels that have been superficially hardened.
B	Steel ball bearing with diameter of 1.16	100	Alloys of copper and aluminum. Soft steels
C	Diamond cone	150	Steel-hard cast iron. Steels that have been deeply hardened
D	Diamond cone	100	Steels that have been moderately hardened and thin steels
E	Steel ball bearing with diameter of 1.8	100	Cast iron. Alloys of Al and Mg. Bearing metals
F	Steel ball bearing with diameter of 1.16	60	Annealed copper alloys
G	Steel ball bearing with diameter of 1.16	150	Phosphor bronze. Copper. Be. Malleable iron
H	Steel ball bearing with diameter of 1.8	60	Al. Pb. Zn

Table 10.
Specifications of Rockwell hardness tests.

until it reaches its elasticity, and stress and strain follow Hooke's law. The coefficient of this equation is called elastic modulus, which depends on the inherent conditions of materials. In the stress–strain diagram, the spot that shows the required stress to begin the plastic deformation, is called the elastic limit and the intended stress is called yield stress. Frangible materials get fracture after this stage, but other materials enter the plastic deformation stage. The increase in force continues as long as the effect of increasing the force is higher, due to hardness work, than effect of reducing the force due to reduction of the cross section, and when the two effects are equal, the force applied to the sample reaches its maximum value and at this point, the object starts to necking from the weakest point. Then increase of force continues until the material reaches a breakpoint and the stretch test is completed. To perform the tension test, the tensile test samples must first be prepared according to ASTM standards. The test consists of the following steps:

- To specify length of the gage on the sample. Length of the part that has the minimum diameter is called gage. Length, width and thickness of the gage are measured with the caliper.

- To place the gage part of the sample in the jaws of the apparatus, to proliferate the relative length and to decrease the cross-section.

- To set the initial status of the apparatus to zero and the speed of the jaws.

- To draw a curve of force variation in terms of length proliferation.

- To measure proliferation of the relative length and to reduce the cross-section after fracture of samples [39, 56, 59].

3. Experimental methodology

3.1 Research objective

Properties such as biocompatibility, corrosion resistance and a high strength ratio to weight have led to the use of titanium and its alloys, in spite of high prices in the pharmaceutical, petrochemical, aerospace industries. In many of these applications, the need for titanium jointing or their alloys is in a variety of ways. One of these jointing methods is welding. One of the most commonly used methods of titanium and its alloys welding is the arc-method of tungsten-neutral gas electrode. The variables of this process have a direct impact on the properties of the final product. Therefore, when welding of these alloys is required in different applications, the choice of welding variables should be achieved in such a way that the final microstructure and desired properties are obtained. In this welding process, variables such as current intensity, welding voltage, welding speed and protective gas have a significant effect on the final properties of the product. The purpose of this study is to investigate the impact of the input heat variety (welding current intensity) on the microstructure and mechanical properties of alloy-welded parts in Ti-6Al-4 V by arc-method of tungsten-neutral gas electrode. For this purpose, welding operation was done by selecting three current intensities of 80, 90 and 100 amperes and the results of the analyzes for these three current intensities were compared.

3.2 Laboratory facilities and raw materials

In this study, tungsten non-consumable Manual Metal, Ti-6Al-4 V sheets, argon-grid gas 6.0 as a protective gas and consumable and non-consumable welding wires of ER Ti.5 and EW Th.2 were used. The used welding machine was the manufacture of the Gam-Electric company, the 400 model. The work-piece was jointed to the positive pole and the electrode was jointed to the negative pole and the polarity of welding was DCEN.

3.3 Method of performing the research

On six sheets of 250 × 125 × 2 mm of Ti-6Al-4 V, three tests of PQR were used with welding process with non-consumable tungsten electrode with neutral gas protection with different welding variables. Then, the samples were joined by GTAW welding method by Square Joint with a number of welds. The welding operation at the first stage was performed at a current intensify of 30 amps, but the weld did not penetrate completely back of the sample and the test failed. At the next stage, in order to achieve full penetration, the sample joint model was changed from Square Joint to V shape, and the back-weld was used to protect the molten pool on back strain gas. In this case, the current intensity increased to 50 amps, and again, no full weld penetration was achieved after welding operation. At the last stage, the current intensity increased to 80A according to the above method, and in this case, full penetration was achieved. PQR test were taken with the current intensity of 90 and 100 amps in the second and third joint samples, and, as in the first case, a full penetration was

Current intensity	**80 amps and 90 and 100**
Voltage range	11–13
Joint model	V shape
Groove Angle	60 degree
Current Rate	Lit/min 25
Consuming Gas	Argon with a purity of 99/9999%
Tungsten Metal diameter	6 mm/1
Welding Wire Diameter	1 mm
Type of welding wire and consumable and non-consumable Welding speeds	ER Ti-5 و EW Th-2
Welding speeds	cm/sec 0.22 (for 80 amps current intensity)
	cm/sec 0.263 (for 90 amps current intensity)
	cm/sec 0.33 (for 100 amps current intensity)
Root Opening	mm 0–1
weld Bead	5–7 mm
Number of weld layers	1 layer
N/A	Pre-heat
N/A	PWHT
Gas current rate	12 lit/min

Table 11.
Specifications and variables of the performed welding processes.

obtained. It should be noted that all tests were performed according to ASME Section IX standard.

Table 11 shows the variables and Specifications of the performed welding operations. Also, the images of the performed welding operations have been in **Figure 9**.

3.4 Study methods and evaluating the properties

After the welding operation, the welding joint site was investigated using an optical microscope after metallography, so that the impact of changing the current intensity is investigated on the resulting microstructure and also their comparison is conducted with each other and with the base metal. In order to investigate the mechanical properties, samples were sent to Razi Foundation for Applied Sciences Laboratory for tests of

- Tensile,
- Impact and
- Hardness and Metallography

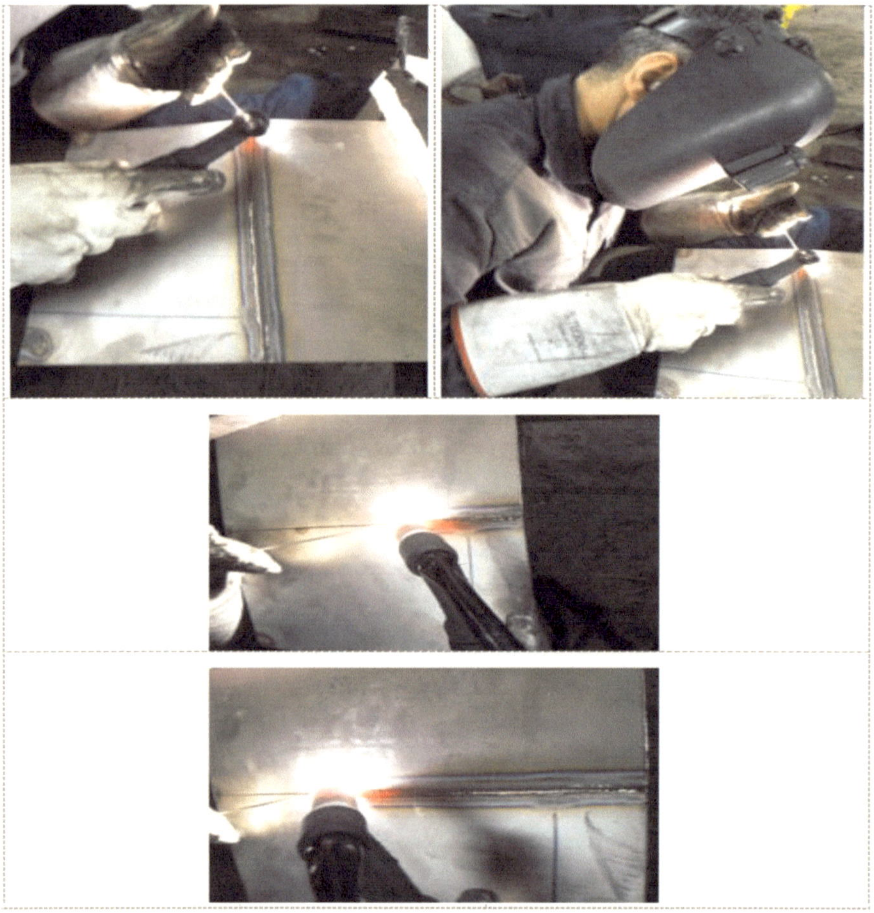

Figure 9.
Images of performed welding operations.

By the center.

In the next chapter, we will present the results of these experiments and will analyze and review them, and these results will be compared with the results of other studies.

4. Results & discussions

4.1 Introduction

This section has analyzed the data and the obtained results from the experiments. As it was already stated, these experiments have been performed on Ti-6Al-4 V. Three different electrical current intensities (welding amps) have been used to conduct experiments that included 80.90 and 100 amps, and the impact of the electric current intensity on the welding specifications has been studied. For this purpose, the weld hardness, and as well as the tensile strength of the samples have been studied and compared with the base electrode samples. In addition, the gained microstructure of the weld has been evaluated and studied in different welding areas and its different phases have been analyzed. In this chapter, the impact of electrical current has been analyzed on mechanical properties. Elongation increase:

Elongation increase has been measured also in this experiment. Elongation increase least rate has been related to the samples that have been jointed with the least current rate (80 amps) and it has been about 30% less than those that have been jointed to each other with current s above 90 and 100 amps. Elongation increase expresses the flexibility in some degree and has an inverse relationship with the weld strength. Its reason for this can be attributed to the weld structure, so that in welds that have been welded with a current of 80 amps, the beta phase has been more precipitated in the welds boundary, which causes the weld flexibility to decrease and Elongation Increase also decreases (**Figure 10**).

4.2 Tensile test

One of the most important welding targets in metals is to create a higher strength, or at least equal to the base metal, which can withstand the forces applied to the sample. In all welds, an attempt is made to increase the strength of the weld above the base metal, and there be no break in the weld and adjacent area of the weld and the

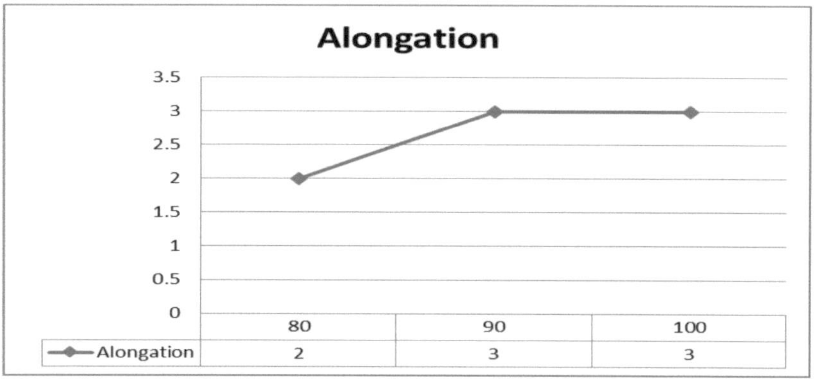

Figure 10.
Current effect on elongation increase.

break is made in the base metal. One of the most important tests for weld quality measurement is tensile test. In this test, the original sample and then other samples have been examined. Each sample has been tested three times for tensile test.

4.2.1 Yield strength

Yield strength for the measured samples is observable in the below figure. As it is known, the highest yield strength belongs to the current of 80 amps, due to the high percentage of beta phase, and as well as the impact of welds size (**Figure 11**).

By increasing the current rate, due to the change in the input heat rate to the weld area and also the change in the speed rate of the electrode, we observe that the yield strength decreases, so that with an increase in the current from 80 to 90 amps, we observe a 4% decrease in Yield Strength, and we will have a 21% decrease in strength by current increase to 100 amps (compared to 80 amps). In short, with welding current increase, we will observe a decrease in Yield Strength.

4.2.2 Ultimate strength

Ultimate Strength of the samples also has been measured also that has been shown in the below figure. As it is known, the highest Ultimate Strength belongs to the current of 80 amps and the lowest strength rate is also related to the current of 100 amps, which can be attributed to how the structure of the welds is changed, which the beta phase that is a needle-shape, it has been formed about 40% of the ground phase in the current of 80 amps, but in the current of 100 amps, about 30% of the ground phase is due to the needle-shape phase of the beta phase, this phase increases the ultimate strength of the weld, and during the 90 amps, 35% of the ground phase belongs to the beta phase. Also, the alpha phase, which somewhat reduces the strength, has the highest rate at 100 amps, with an approximate ratio of 70%, but within 80 Amps, only 60% of the phase forms the alpha phase. This structure change affects the ultimate strength of the weld (**Figures 12** and **13**).

Generally, the current increase to 100 amps has improved the welding conditions and has made it possible to achieve the lowest strength rate, which is above 1000 mega-Pascal, that it is because of the reduction of the input thermal energy to the weld area, due to the electrical current, which causes improvement of the localized melting

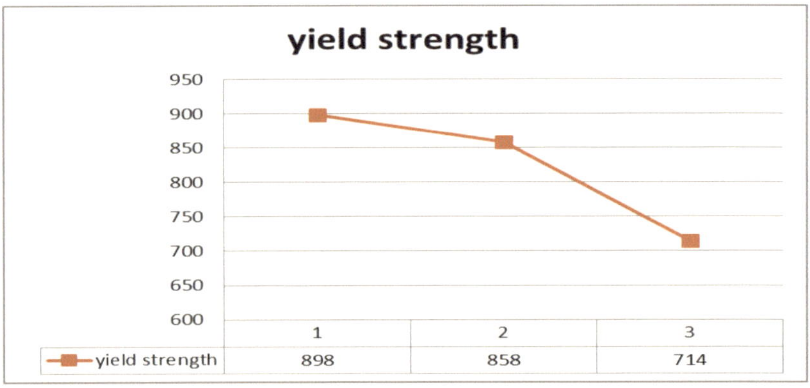

Figure 11.
Yield strength of samples that have been welded with electrical current intensities of 80, 90 and 100 amps.

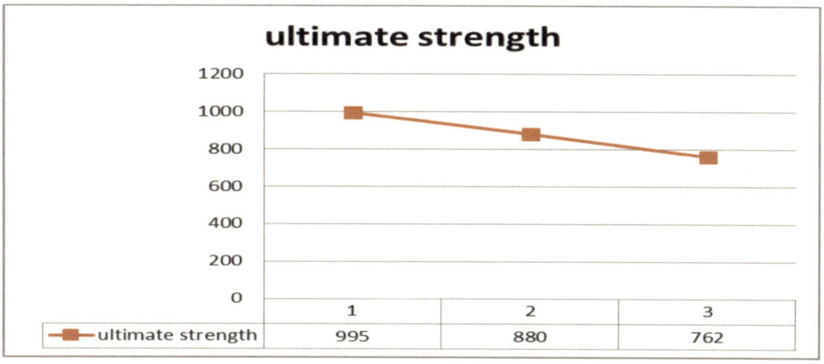

Figure 12.
Ultimate tensile strength of samples that have been welded with 80, 90 and 100 amps.

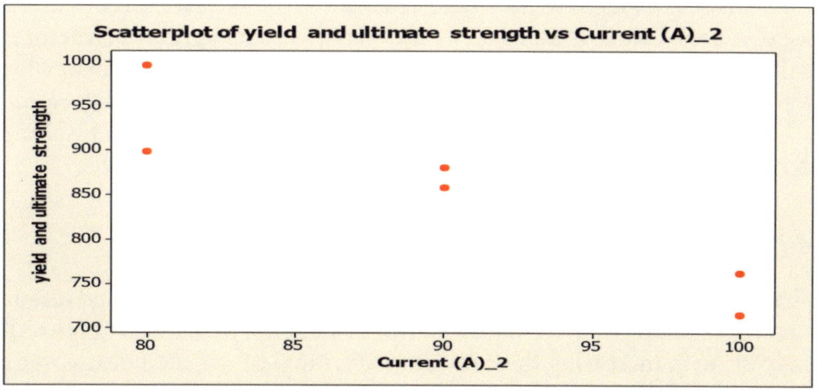

Figure 13.
Comparison of the ultimate strength and the tensile yield strength of the samples, which have been welded with electrical currents intensity at 80, 90 and 100 amps.

phenomenon and the melting of the two edges of the weld occurs better, which increases the flexibility of the welding.

In fact, the reason for increasing strength can be attributed to the weld microstructure, and the microscopic structure indicates that the weld area includes the alpha phase coarse-grain, and granular (bright) and the grain boundary phase of beta (dark). The beta phase stagnation in the alpha phase causes locking and the strength increase, but at a 100-amps the needle density (beta phase in alpha), the current ratio of 80 and 90 amps has been lower, which causes the strength of these samples reduces in comparison to two samples. The transformation of the base metal needle-shape structure (which increases the strength), into a coarse-weld structure is the main factor in affecting the mechanical specifications.

4.2.3 Micro-hardness

Vickers Micro-hardness test has been used to measure the hardness of the samples. Hardness test results show that increasing the electrical current reduces the weld hardness. As the electrical current increase causes the temperature increase and because of that the gradient beta phase has been decreased and the mechanical properties of the samples have been changed and the mechanical specifications of the samples have been changed and the increase of the electrical current causes the fine-weld structure in the weld area, which increases the hardness of the material.

However, the three samples tested showed different harnesses, but this hardness was tolerable in a reasonable range, so that the difference is less than 2% of the average hardness of the samples, due to the lack of uniformity of the phases.

4.2.4 Hardness test (Vickers) for current of 80 amps

Measuring samples that have been welded with a current of 80 amps, show that they have a higher hardness than the original metal sample (**Figure 14**).

4.2.5 Hardness test (Vickers) for current of 90 amps

The reason for the low hardness of samples that have been welded with a current of 90 amps can be attributed to the microstructure of the samples. In the current of 90 amps, we observe a coarse-grained structure that this coarse-grained structure reduces the hardness. The least difficulty related to the heat affected area, which is due to the effect of the annealing heat, due to the process which makes the area soften. In fact, the input heat rate in the affected area increases the grain size, resulting in the strength reduction and ultimately, the hardness reduces too (**Figure 15**).

4.2.6 Hardness test (Vickers) for 100 amps

The hardness of the samples have been welded at 100A has been measured and shows that these samples have a higher hardness than the base metal (**Figure 16**).

In this test, with increasing the current to 80 amps, we see an increase in the weld hardness, but the hardness decreases during 90 amps and eventually the hardness increases to reach the highest level by increasing the current to 100 amps (**Figure 17**).

The size and morphology of these grains depend on the heat transfer during freezing. The first determinant of grain size is the weld input heat. In this way, the grain size will be larger if using higher output heat. The mechanical properties of the melting area of titanium, depend on how the phase state changes during cooling in the phase stability temperature range, in addition to the initial grain size. The ultimate microstructure depends on the cooling speed above the state change temperature, which itself is a function of the welding process type, process parameters, and other welding conditions, such as the geometric shape of the piece and the method of

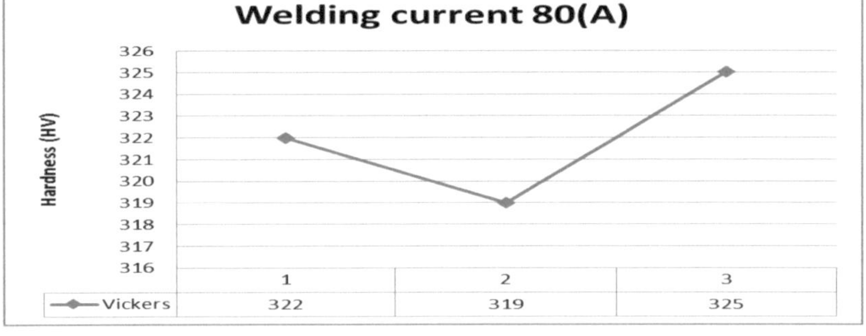

Figure 14.
Comparison of the hardness of the samples that have been jointed with the electrical current intensity of 80 amps.

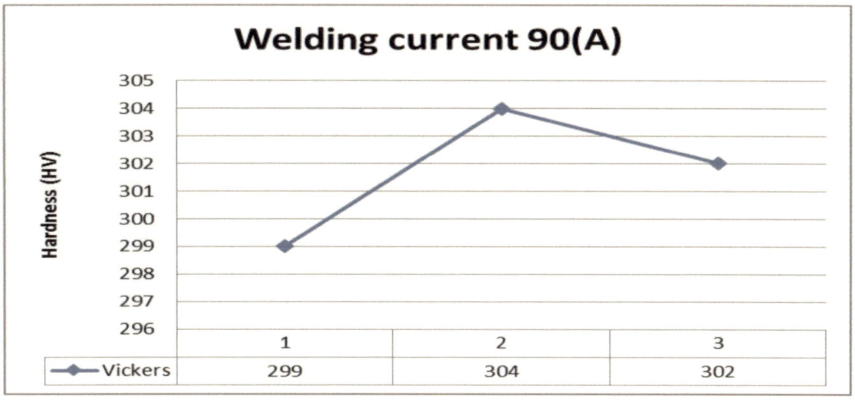

Figure 15.
Comparison of the hardness of the samples which have been jointed with the electrical current intensity of 90 amps.

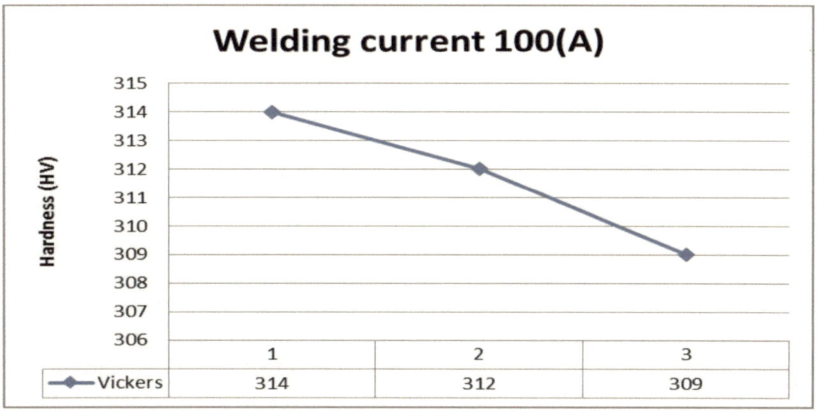

Figure 16.
Comparison of the hardness of the samples which have been jointed with the electrical current intensity of 100amps.

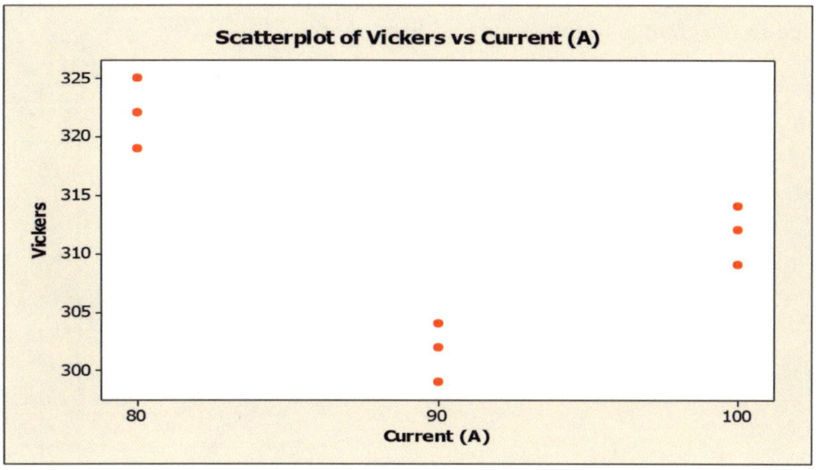

Figure 17.
Comparison of the hardness of samples that have been jointed with different electrical current.

fastening the piece. The microstructure is very fine and needle-shape, and its mechanical properties are strength and hardness versus low flexibility. The microscopic structure indicates that the weld area includes the alpha phase coarse-grain, and granular (bright) and the grain boundary phase of beta (dark). The formation of a base metal needle-shape structure that causes increasing the hardness to the base metal, and as well as coarseness of grains, is also responsible for affecting on mechanical properties such as hardness. However, in 100 A, the needle density (beta phase in alpha) has been lower than 80 and 90 A, which causes the reduction of the hardness of these samples than the other two. The formation of a base metal needle-shape structure, which causes the increase of the hardness to the base metal, and as well as coarseness of grains, is also a factor in affecting on the mechanical properties such as hardness.

4.2.7 Impact test: Charpy test

One of the tests that indicates the toughness of the material is the impact test. This test shows the weld energy level. One of the characteristics of titanium is toughness and malleability, which makes it absorb the energy of the hits and the forces and is not broken like a cast iron and Regarding that a good weld should be like the base metal, that does not cause a change in the structure and mechanical properties, so a good weld should also retain the toughness. The results of Charpy test for 80, 90, and 100 amps indicate that the toughness of the weld increases, by the electrical current increase and more energy is in need to break the weld. Titanium alloys contain one or more phase stabilizing elements, and their welding can effectively change the strength, flexibility and toughness of the weld-metal and the adjacent area of the weld. Usually, if these alloys contain more than 20% of the beta phase, their weldability is considered to be weak. Titanium beta alloys contain sufficient amounts of beta phase stabilizing elements.

This group of titanium alloys can be welded, but those beta-type titanium alloys that contain high amounts of phase-stabilizing elements have poor weldability because the welding metal has a high tensile strength. The mechanical properties of the welding zone, including the toughness of the titanium alloys, depend on the microstructures of FZ and HAZ areas, where the results of the Charpy test for 80, 90 and 100 amps indicate that with increasing electric current, the toughness of the welding increases and more energy is needed to break the weld, that its reason can be attributed to the change in the alpha and beta phases.

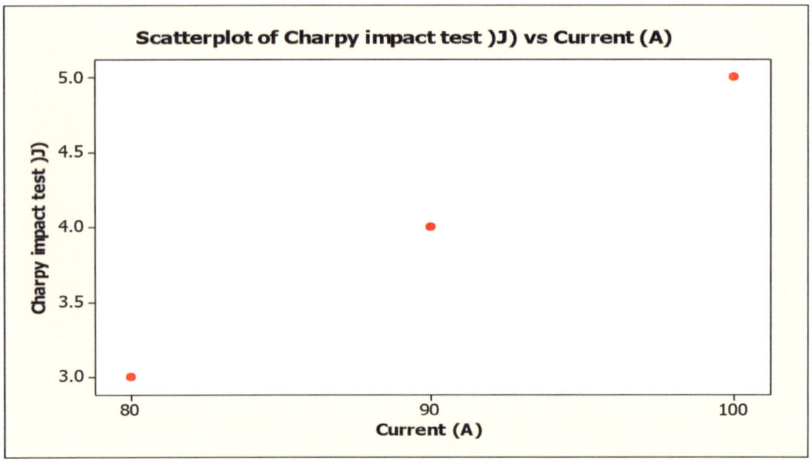

4.2.8 Welding section

The weld cross-section is semi-elliptic (pelvis-like), in which three distinct areas can be identified. Base metal area, weld area (pelvis-like) and heat affected area (**Figure 18**).

The area of the welded area and the heat affected area are directly affected by the welding parameters, and the welding parameters make this area smaller or larger. It is more appropriate that the welding area be smaller to give a good look to the piece and it will be suitable if the strength is appropriate.

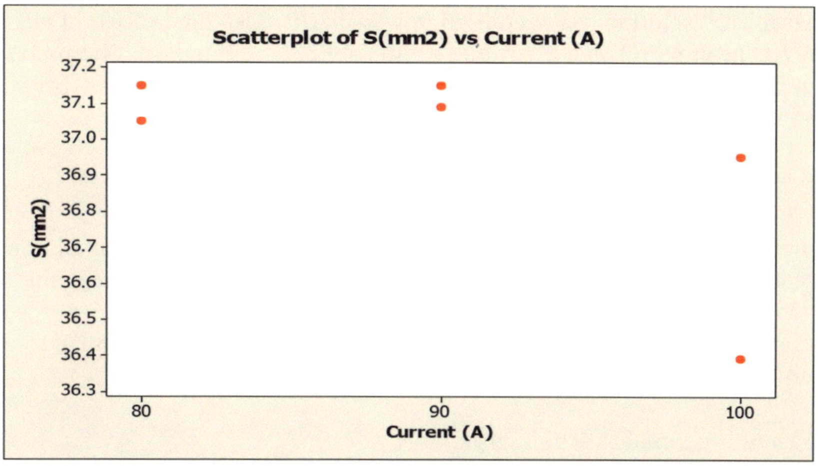

The area of the welded area and the heat affected area directly are affected by the welding parameters. The expansion of the weld area to the end of the metal will have a great effect on the quality and strength of the weld. For better examination, the weld is cut transversely, so that the cross-section can be clearly seen. In samples have been welded with less than 60 amps electrical current, it is clear that the lower area of the pieces is not well welded to each other, due to the lack of penetration of the tool into the lower part of the pieces, which it caused the pieces have not been fully jointed. The low electric current has been caused the metal in the two lower parts of the screen cannot be well-drained, which this will reduce the ultimate strength of the weld. The

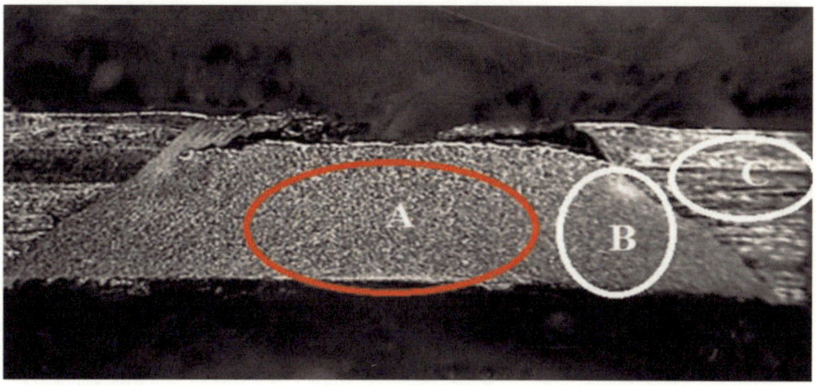

Figure 18.
Different areas of welding a: Welding area and mixing B: Heat affected area C: Base metal.

higher the electrical current, more heat will be produced, therefore the area of these two areas extends.

4.3 Metallography

For metallography of the weld samples, first they have been cut transversal, and the cut surface has been polished by the polish machine. Standardization of metallography has been carried out using ASTM E3–11 standard. After polishing the samples, using the etching process, samples have been prepared to see below under the microscope. For this purpose, the etching process has been used according to ASTM E407–2015 standard. Kroll solution has been used to visualize it. Also the process of optical microscopy has been performed according to ASTM E883–11 standard. Microscopic structure in base metal includes the alpha phase coarse-grain, and granular (bright) and the grain boundary phase of beta (dark).

4.3.1 Microstructure in base phase

The structure of the base metal is more regular and fine grain. In all samples, the alpha phase has been dominant, and the beta phase is observed only in the grain boundary (**Figure 19**).

Microstructure in the weld area includes the alpha phase fine-grain, and granular (bright) and the grain boundary phase of beta (dark).

4.3.2 Microscopic structure in the weld area

The fine grain structure in the weld area affects both the hardness and the tensile strength in accordance with Hall's hypothesis. It shows the microstructure of the affected area by heat, which is the same coarse-grain and needle-like jagged microstructure. Partial melted grains in the common solid-molten area are suitable sites for solid phase growth into the molten pools. The grains are increased in as the surface from the common solid-molten area to the weld line-center. Because the mechanical properties of the weld, especially its flexibility, depend on the grain size (**Figure 20**).

4.3.3 Microstructure due to weld

The following figure shows that the microstructure of the weld area is fine-grained, needle-shape and jagged. Titanium at medium temperature and pressure has

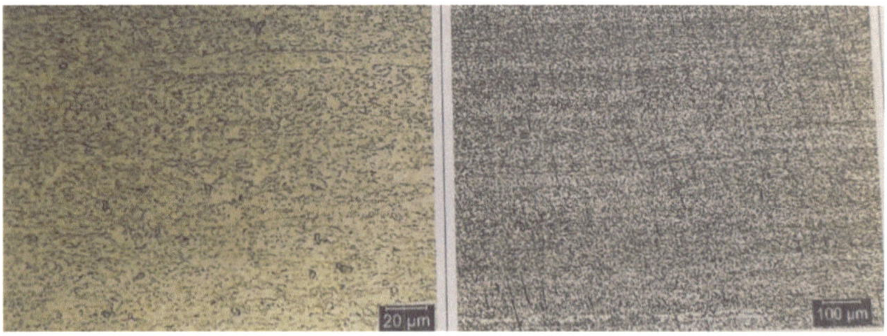

Figure 19.
Basic metal microstructure with two magnifications for 80 amps.

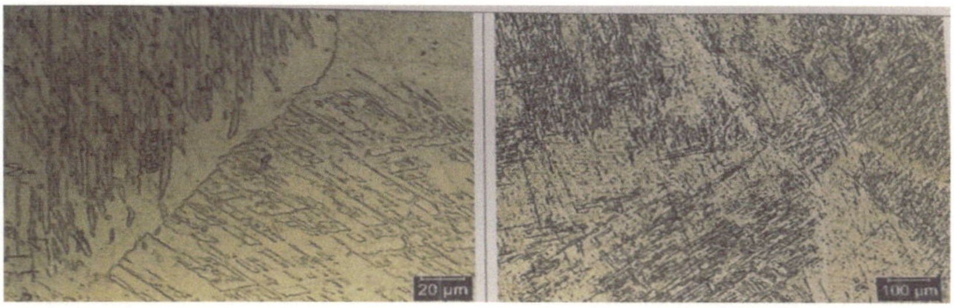

Figure 20.
The microstructure of the heat affected area with two magnifications for 80 amps.

a HCP crystalline structure with a compression ratio of (c/a) 1.587 (α phase). At 890 degree, Titanium is transformed into a β-phase with the crystalline BCC structure by the allotropic transformation, which this structure remains stable until the melting point (1678 degree). An increase in the electrical current increases the temperature during the process, which due to that, beta grain boundary phase has begun to grow and is increasing, and this increase in the beta phase in the weld area is higher than the heat affected area, due to the higher thermal gradient. Another factor influencing the beta phase increase is the cooling rate, which is highly dependent on the linear speed of the tool and the rotational speed of the tool has a minor effect, but it also can be ignored the effect of the cooling rate on the grain size of the beta phase and its amount, due to the linear speed constant in the disorder friction welding process of all five samples (**Figure 21**).

4.4 Material microstructure evaluation

During the welding operation, the microstructure of the weld area and the affected areas by it change, so evaluating these changes can help you understand the effects of the parameters on the mechanical and physical properties of the weld. For this purpose, the samples have been metallographed and then have been examined by optical microscopy. First, the microstructure of the base metal has been evaluated. The two phases of alpha and beta are the main phases of this alloy that can be clearly seen. The bright areas are related to the alpha phase, and the dark areas are also related to the beta phase (it should be noted that these two phases consist of two different crystalline structures). As it is evident in the figure, these two phases are completely

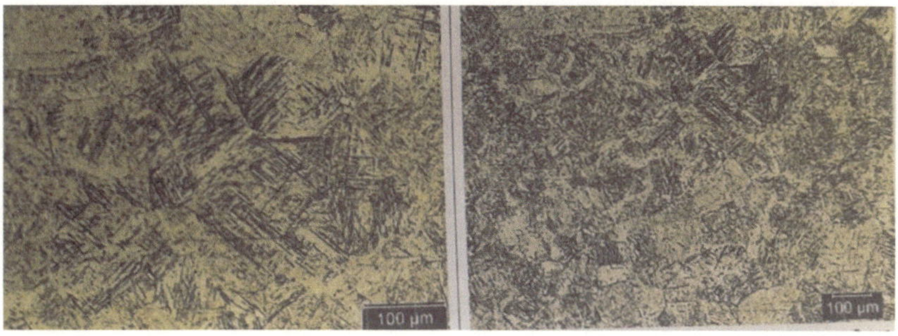

Figure 21.
Welded area microstructure with two magnifications for 80 amps.

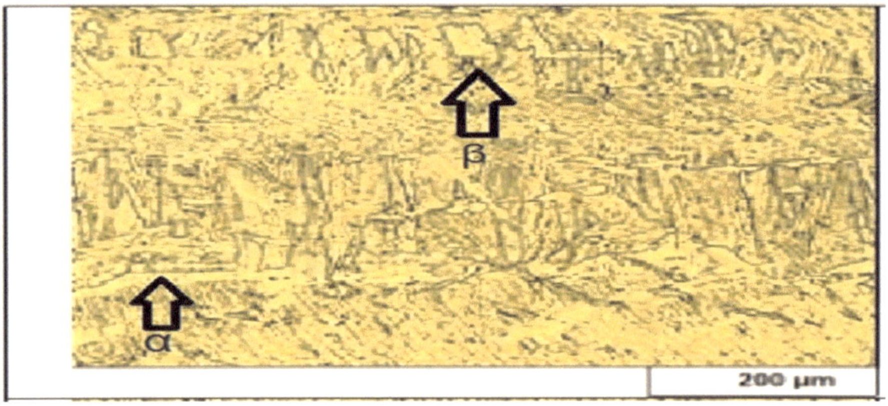

Figure 22.
Base material microstructure.

dispersed in each other, and the dominant phase is the alpha phase that the beta phase has been dispersed inside it (**Figure 22**).

4.5 Heat affected area

One of the areas that greatly affects the quality of the weld is the area affected by heat, which the microstructure of the area specifies the site and the type of weld break. The cold and stronger area around the weld area resists against the form that makes the affected area by thermo-mechanical operation. The specifications of the made form change in this area may also be covered by subsequent fuzzy transformations due to thermal cycles occurring during the process. Microscopic images show that the structure of the base metal, which is coarse-grain, has been turned into a more fine grain structure (of course relatively) in the heat affected area. Of course, this effect is less visible in samples that have been jointed in a low current. For example, in samples that have been jointed in 80 amps, also it is seen the structure of the alpha field along with the beta phase and is similar to the main metal, and this structure is more similar to the base metal, but we see the grains coarser and decrease of the beta6phase with an increase in the intensity of the current to 90 amps and clearly coarseness of grains is visible in 100 amps (**Figure 23**).

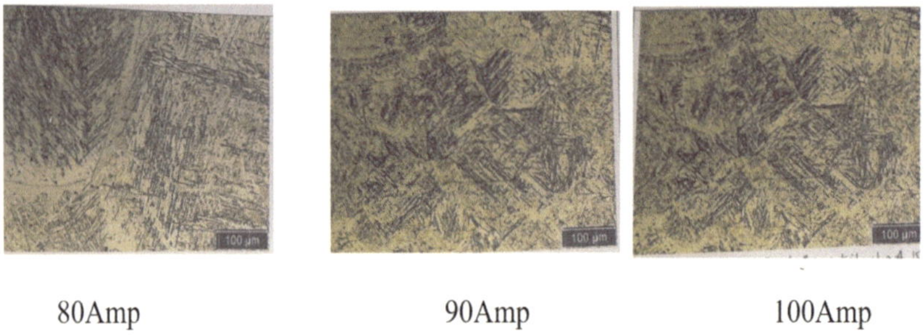

Figure 23.
Heat affected area microstructure in three different samples: A(80 Amp), B(90 Amp), C(100 Amp).

4.6 Weld area

Weld area is the most important Weld area that determines the quality of the weld. All the intermixing and thermo-dynamic phenomena happen in this part, so the microstructure of this area is very important. The special specifications of Titanium weld area are the presence of coarse grains during solidification of the weld metal in the molten pool area, resulting in a sharp decrease of its flexibility. The microstructure have been jointed in the samples with the least electrical current that they show the needle structure of the base metal has been converted into the coarse-grained structure in the samples, and as the electrical current increases, the size of the grains is also increased, so that the size of the grain in the metal weld area which has been jointed to with 100 amps is greater than the size of the grain of the weld area which has been jointed with 80 amps and the density of the needle-shape of the samples has been also increased.

This increase in density of the needle-shape (the placement of the beta phase in the alpha phase) affects the mechanical properties of the samples, which has been increased the strength of the samples compared to the base metal samples. The transformation of the needle-shape structure of the base metal (which increases the strength), into a coarse-grained structure is the main factor in effect on the mechanical properties. In fact, in the low currents, the made heat softens the weld (**Figure 24**).

The weld structure in samples that have been jointed with high-current shows that these structures contain a large number of needle-shape grains, which this is far more than the base metal.

Needle-shape grains behave like dislocation and prevent the movement of the material in the tensile test process and cause the weld strength increases compared to the main metal and break does not occur in the weld area. This mode is an ideal weld. But the interesting point is being needle-like density in a sample has been welded with 90 A, which being needle-shape density was reduced with 80 A and 100 A, which reduces its mechanical properties, which is also confirmed by tensile and hardness tests.

In the molten welding, titanium alloys, non-alloy titanium and alpha titanium alloys have a good weldability. That is, there is no significant difference in terms of microstructure and mechanical properties between the weld areas, the adjacent to the weld area and the base metal, and the resulting weld has sufficient strength with suitable flexibility.

4.7 Experiment results

4.7.1 Effect of electric current on the weld hardness

The results of the experiments show that the hardness of the base metal is always more than the hardness of the weld, which can be attributed to the softening and

80Amp 90Amp 100Amp

Figure 24.
Weld area microstructure in three different samples: A(80 Amp), B(90 Amp), C(100 Amp).

coarse structure of the metal in the weld area, where the output heat only results in the coarse-grain of the weld area and, as a result, increased hardness. Of course, with increasing the rate of electrical current from 80 amps to 100 amps, we will face the reduction of the hardness, because the intensity increase in the electrical current will reduce the heat to the welded area and the intermetallic compounds will not be formed that will affect the hardness. Therefore, it can be expected that by increasing the electrical current above this rate, we can observe the reduction of the hardness of the weld above the base metal. It should be noted, however, that during 90 amps, the hardness decreases that can be related to the alpha phase intensity and the beta phase.

4.7.2 Effect of electrical current intensity on strength

Samples have been tested by tensile test to obtain the tensile strength of the samples to investigate the effect of electrical current intensity on the weld quality. The results of the experiments show that samples that have been welded with a current rate of 80 and 90 amps have a higher strength than the base metal and break does not occur in the weld. Of course, the samples that have been welded with an electrical current of 100 amps have a lower strength than the two previous samples, but this experiment showed that the lower level of the electrical current rate that causes the weld does not have the required quality is below 80 amps (according to other parameters of this experiment).

4.7.3 Effect of welding current intensity on microstructure

Microscopic results show that the structure of the base metal consists of coarse-grained alpha-beta phases that have been interpenetrated in needle-shape form, which this structure is converted to a coarse-grain structure with a needle-shape low density in high electrical current intensity in the weld area, which is one of the main reasons for reducing the weld strength which causes no barrier (grain boundary) against material flux and consequently reduces the strength of the material, so that in samples that have been jointed with higher electrical current, we observe the lower hardness and strength.

4.7.4 Offers

Considering the great capabilities of titanium welding and the gained experiences in this research, suggestions are suggested for improving this method:

1-Performing the process inside the neutral gas:

Due to the high impact of titanium on the environment and the possibility of entering nitrogen, hydrogen and oxygen gases into the weld pool, it is suggested that this process is also carried out in a vacuum environment to increase the weld quality.

Chapter four Appendices figures: (Welding Procedure Specification).

A. Appendix

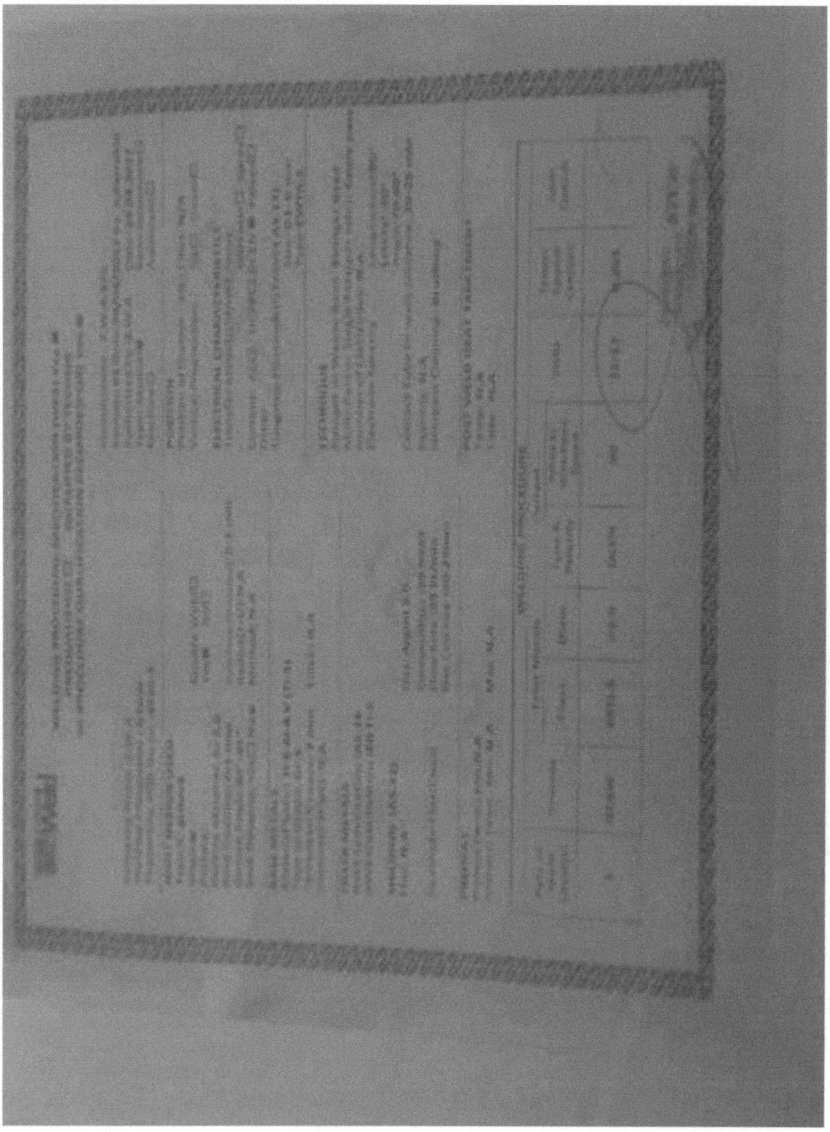

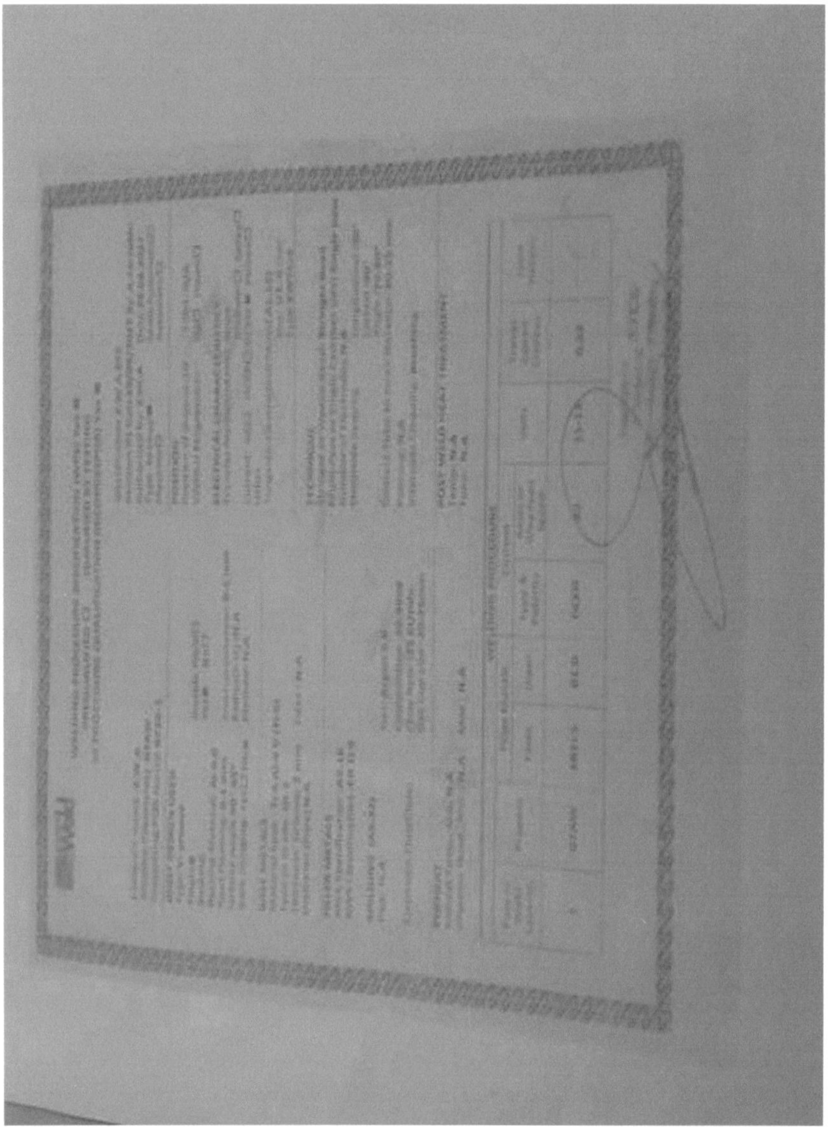

DOI: http://dx.doi.org/10.5772/intechopen.106677

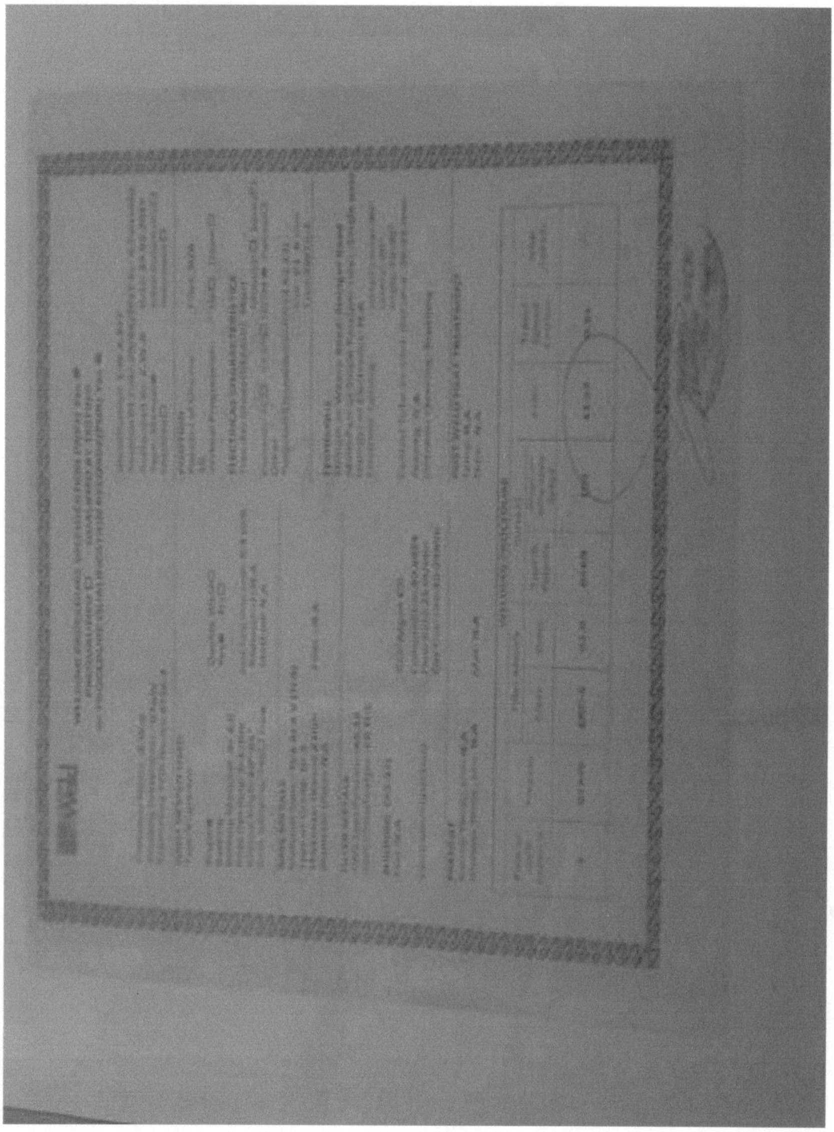

Author details

Alireza Farrokhi
National Valley University in California, USA

*Address all correspondence to: alifarrokhi5@gmail.com

References

[1] Leyens C, Peters M, editors. Titanium and Titanium Alloys: Fundamentals and Applications. Germany: John Wiley & Sons; 2003

[2] Welsch G, Boyer R, Collings EW, (Eds.). Materials Properties Handbook: Titanium Alloys. USA: ASM International; 1993

[3] Navedra Navedtra 14251 A Cost Comparison of welding processes.

[4] Lütjering G, Williams JC. Titanium. Berlin/Heidelberg, Germany: Springer Science & Business Media; 2007

[5] Boyer RR, Briggs RD. The use of β titanium alloys in the aerospace industry. Journal of Materials Engineering and Performance. 2005;**14**(6):681-685

[6] Boyer RR. An overview on the use of titanium in the aerospace industry. Materials Science and Engineering: A. 1996;**213**(1–2):103-114

[7] Charles C. Microstructure Model for Ti-6Al-4V Used in Simulation of Additive Manufacturing [thesis]. Sweden: Luleå tekniska universitet; 2016

[8] Donachie MJ. Titanium: a technical guide. USA: ASM International; 2000

[9] Chen JC, Pan CX. Welding of Ti-6Al-4V alloy using dynamically controlled plasma arc welding process. Transactions of Nonferrous Metals Society of China. 2011;**21**(7):1506-1512

[10] Niagaj J. Peculiarities of A-TIG welding of titanium and its alloys. Archives of Metallurgy and Materials. 2012;**57**(1):39-44

[11] Babu NK, Raman SGS. Influence of current pulsing on microstructure and mechanical properties of Ti–6Al–4V TIG weldments. Science and Technology of Welding and Joining. 2006;**11**(4):442-447

[12] da Cunha TV, Bohórquez CEN. Ultrasound in arc welding: A review. Ultrasonics. 2015;**56**:201-209

[13] Singh AK, Dey V, Rai RN. Techniques to improve weld penetration in TIG welding (a review). Materials Today: Proceedings. 2017;**4**(2):1252-1259

[14] Yang M, Zheng H, Qi B, Yang Z. Effect of arc behavior on Ti-6Al-4V welds during high frequency pulsed arc welding. Journal of Materials Processing Technology. 2017;**243**:9-15

[15] Kumar P, Datta CK. Pulsed parameters optimization of GTAW process for mechanical properties of Ti-6Al-4V alloy using Taguchi method. IJESIT. 2012;**1**(1):75-80

[16] Akbari M, Saedodin S, Toghraie D, Shoja-Razavi R, Kowsari F. Experimental and numerical investigation of temperature distribution and melt pool geometry during pulsed laser welding of Ti6Al4V alloy. Optics & Laser Technology. 2014;**59**:52-59

[17] Stenbacka N, Choquet I, Hurtig K. Review of arc efficiency values for gas tungsten arc welding. In: IIW Commission IV-XII-SG212, Intermediate Meeting, BAM, Berlin, Germany, 18–20 April, 2012. Berlin, Germany; 2012. pp. 1-21

[18] Tolvanen S. Microstructure and Mechanical Properties of Ti-6Al-4V Welds Produced with Different Processes [Thesis]. Sweden: Halmers University of Technology; 2016

[19] Taylor JC, Hondrum SO, Prasad A, Brodersend CA. Effects of joint configuration for the arc welding of cast Ti-6Al-4V alloy rods in argon. The Journal of Prosthetic Dentistry. 1998; **79**(3):291-297

[20] ASM International. Handbook Committee., ASM Handbook. Vol. 6a. USA: ASM International; 1993

[21] Weman K. Welding Processes Handbook. Sweden: Elsevier; 2011

[22] Leary RK, Merson E, Birmingham K, Harvey D, Brydson R. Microstructural and microtextural analysis of InterPulse GTCAW welds in Cp-Ti and Ti–6Al–4V. Materials Science and Engineering: A. 2010;**527**(29):7694-7705

[23] Wells ME, Richardson J. Recommendations and Requirements for Welding and Inspection of Titanium Piping for US Navy Surface Ship Applications (No. CARDIVNSWC-TR-61-98-18). Bethesda, MD: Naval Surface Warfare Center Carderock Division; 1999 Survivability Structures and Materials Directorate

[24] Khaled Z. An investigation of pore cracking in titanium welds. Journal of Materials Engineering and Performance. 1994;**3**(3):419-434

[25] Karimzadeh F, Salehi M, Saatchi A, Meratian M. Effect of microplasma arc welding process parameters on grain growth and porosity distribution of thin sheet Ti6Al4V alloy weldment. Materials and Manufacturing Processes. 2005; **20**(2):205-219

[26] Panwariya C. Evaluation of Nitride Layer on Ti-6Al-4V Titanium Alloy by Tungsten Inert Gas (TIG) Nitriding Method (thesis). Rourkela, Odisha: NIT; 2016

[27] Semiatin SL, Knisley SL, Fagin PN, Barker DR, Zhang F. Microstructure evolution during alpha-beta heat treatment of Ti-6Al-4V. Metallurgical and Materials Transactions A. 2003; **34**(10):2377-2386

[28] Ucok I, Kramer LS, Gungor MN, Wolfe P, Dong H, Tack WT. Effect of welding on microstructure and tensile properties of flowformed Ti-6Al-4V tubes. Materials Science and Engineering: A. 2005;**410**:160-164

[29] Radhakrishnan VM. Welding Technology and Design. New Dehli, India: New Age International; 2006

[30] Mehdi B, Badji R, Ji V, Allili B, Bradai D, Deschaux-Beaume F, et al. Microstructure and residual stresses in Ti-6Al-4V alloy pulsed and unpulsed TIG welds. Journal of Materials Processing Technology. 2016;**231**:441-448

[31] Balasubramanian M, Jayabalan V, Balasubramanian V. Developing mathematical models to predict tensile properties of pulsed current gas tungsten arc welded Ti–6Al–4V alloy. Materials & Design. 2008;**29**(1):92-97

[32] Balasubramanian TS, Balasubramanian V, Manickam MM. Fatigue crack growth behaviour of gas tungsten arc, electron beam and laser beam welded Ti–6Al–4V alloy. Materials & Design. 2011;**32**(8): 4509-4520

[33] American Welding Society. Specification for Tungsten and Tungsten Alloy Electrodes for Arc Welding and Cutting. USA: ANSI-AWS A5.12M-98, February 1998

[34] American Welding Society. Filler Metal Procurement Guidelines. USA: ANSI/AWS A5.01–93; 1993

[35] Cary HB. Modern Welding Technology. USA: Prentice-Hall; 1979

[36] Albrecht AK. System Development for Analysis of Gas Tungsten Arc Welding [thesis]. USA: Vanderbilt University Institutional Repository; 2005

[37] Saitta Alexander M, Oates WR. Welding Handbook.: Materials and Applications. USA: American Welding Society; 1998

[38] Pires JN, Loureiro A, Bölmsjo G. Welding Robots: Technology, System Issues and Application. Portugal: Springer Science & Business Media; 2006

[39] Kou S. Welding metallurgy. New York, USA: John Wiley & Sons; 2003

[40] Nandagopal K, Kailasanathan C. Analysis of mechanical properties and optimization of gas tungsten arc welding (GTAW) parameters on dissimilar metal titanium (6Al 4V) and aluminium 7075 by Taguchi and ANOVA techniques. Journal of Alloys and Compounds. 2016; **682**:503-516

[41] Babu NK, Raman SGS, Mythili R, Saroja S. Correlation of microstructure with mechanical properties of TIG weldments of Ti–6Al–4V made with and without current pulsing. Materials Characterization. 2007;**58**(7):581-587

[42] Murthy KK, Potluri NB, Sundaresan S. Fusion zone microstructure and fatigue crack growth behaviour in Ti-6Al-4V alloy weldments. Materials Science and Technology. 1997; **13**(6):503-510

[43] Short AB. Gas tungsten arc welding of α+ β titanium alloys: A review. Materials Science and Technology. 2009; **25**(3):309-324

[44] Keshava Murthy K, Sundaresan S. Fracture toughness of Ti-6Al-4V after welding and postweld heat treatment. Welding Journal. 1997;**76**(2):81-s

[45] Mohandas T, Srinivas M, Kutumbarao VV. Effect of post-weld heat treatment on fracture toughness and fatigue crack growth behaviour of electron beam welds of a titanium (a+ b) alloy. Fatigue and Fracture of Engineering Materials and Structures. 2000;**23**(1):33-38

[46] Sundaresan S, Ram GJ, Reddy GM. Microstructural refinement of weld fusion zones in α–β titanium alloys using pulsed current welding. Materials Science and Engineering: A. 1999;**262**(1): 88-100

[47] Salama MM. Fatigue crack growth behavior of titanium alloy Ti-6Al-4V and weldment. TRANSACTIONS-AMERICAN SOCIETY OF MECHANICAL ENGINEERS JOURNAL OF OFFSHORE MECHANICS AND ARCTIC ENGINEERING. 2001;**123**(3): 141-146

[48] Atapour M, Fathi MH, Shamanian M. Corrosion behavior of Ti–6Al–4V alloy weldment in hydrochloric acid. Materials and Corrosion. 2012; **63**(2):134-139

[49] Yang Z, Qi B, Cong B, Liu F, Yang M. Microstructure, tensile properties of Ti-6Al-4V by ultra-high pulse frequency GTAW with low duty cycle. Journal of Materials Processing Technology. 2015;**243**:37-47

[50] Li QM, Wang XH, Zou ZD, Jun W. Effect of activating flux on arc shape and arc voltage in tungsten inert gas welding. Transactions of Nonferrous Metals Society of China. 2007;**17**(3): 486-490

[51] Zhang WJ, Liu YK, Wang X, Zhang YM. Characterization of three-dimensional weld pool surface in GTAW. Welding Journal. 2012;**91**(7): 195s-203s

[52] Onsoien M, Peters R, Olson DL, Liu S. Effect of hydrogen in an argon GTAW shielding gas: Characteristics and bead morphology. In: AWS 75th Annual Meeting, Held April 10–14, 1995. Philadelphia, PA; 1995

[53] Bitharas I, Campbell SW, Galloway AM, McPherson NA, Moore AJ. Visualisation of alternating shielding gas flow in GTAW. Materials & Design. 2016;**91**:424-431

[54] Zhou W, Chew KG. Effect of welding on impact toughness of butt-joints in a titanium alloy. Materials Science and Engineering: A. 2003; **347**(1):180-185

[55] Balasubramanian TS, Balakrishnan M, Balasubramanian V, Manickam MM. Influence of welding processes on microstructure, tensile and impact properties of Ti-6Al-4V alloy joints. Transactions of Nonferrous Metals Society of China. 2011;**17**(6): 1253-1262

[56] Reed-Hill RE, Abbaschian R. Physical Metallurgy Principles. Monterey, Calif, USA: Brooks/Cole Engineering Division; 1973

[57] Raghavan V. Physical Metallurgy: Principles and Practice. New Delhi, India: PHI Learning Pvt. Ltd; 2015

[58] Pearson WB. A Handbook of Lattice Spacings and Structures of Metals and Alloys: International Series of Monographs on Metal Physics and Physical Metallurgy. Vol. 4. London, Paris, New York , Frankfurt: Elsevier; 2013

[59] Dieter GE, Bacon DJ. Mechanical Metallurgy. Vol. 3. New York: McGraw-hill; 1986

[60] Hertzberg RW. Deformation and Fracture Mechanics of Engineering Materials. Hoboken, New Jersey: Wiley; 1989

Section 5

Sustainable Use of Titanium Alloys

Chapter 7

Use of the Advantages of Titanium in the Metal: Organic Framework

Behnaz Delaram, Faezeh Mokari, Mostafa Gholizadeh and Yahya Absalan

Abstract

Titanium is one of the most attractive elements, due to its unique advantages such as stability, recyclability, activity under light absorption, cheapness, and safety. The special characteristics of titanium include different oxidation states, high coordination number of Ti^{4+}, and the ability to form strong bonds with oxygen and different ligands, making it a good candidate for the construction of the new composite named metal–organic framework or briefly MOF. MOFs are composites that have opened a new window toward the scientific world due to their special structure that makes them have some properties, including the highest surface activity, high porosity, tunable pore, and high flexibility in design that make them useful in different applications, such as gas storage and separation, liquid separation and purification, electrochemical energy storage, catalysis, and sensing. Titanium, due to the mentioned properties, has been used as a node in the structure of different MOFs and applied in different fields.

Keywords: titanium, MOF, photocatalyst, modification

1. Introduction

Coordination polymers (CPs) are a class of compounds formed by metal ions (or clusters) coordinated to multidentate organic and/or inorganic ligands. Metal–organic frameworks (MOFs) are a special group of CPs with porous and crystalline one-, two- or three-dimensional structures. These compounds are composed of metal ions or well-defined secondary building units and organic linkers that are connected by coordination bonds and other weak interactions or noncovalent bonds, such as H-bonds, π-electron stacking, or van der Waals interactions [1]. In the 1960s, the MOF kind of structure was firstly introduced by Tomic and others [2–4]. In the 1990s, research on MOF was developed by Robson et al. [5, 6] and later by Yaghi et al. [7].

After the discovery, synthesis, and characterization of the first known MOF structure, MOF-5 ($[Zn_4 (BDC)_3] (DMF)_x$) by Yaghi and coworkers in 1999, a rapid advancement has been made in the chemistry of MOFs [8]. The increasing interest in MOF research can be attributed to their amazing properties, including high specific surface area, ultra-high porosity, tunable surface chemistry, functional diversity, and crystallinity [9].

The main goal of this chapter is the use of titanium as a node in the design of metal–organic frameworks. To achieve this goal, we will specifically investigate the characteristics and properties of titanium as a component in the design of metal–organic frameworks. In the following, considering the features and advantages mentioned in the text for titanium, we will introduce the types of titanium-based MOFs, synthetic methods, and applications. Eventually, the comparison of the properties and performance of titanium-based MOFs with other metal–organic frameworks is discussed.

2. Synthesis of MOFs

MOFs are constructed by connecting inorganic and organic building units through coordination bonds. The inorganic units can be monovalent (Cu^{+}, Ag^{+}, etc.), divalent (Mg^{2+}, Mn^{2+}, Fe^{2+}, Co^{2+}, Ni^{2+}, Cu^{2+}, Zn^{2+}, Cd^{2+}, etc.), trivalent (Al^{3+}, Sc^{3+}, V^{3+}, Cr^{3+}, Fe^{3+}, Ga^{3+}, In^{3+}, lanthanides^{3+}, etc.), or tetravalent (Ti^{4+}, Zr^{4+}, Hf^{4+}, Ce^{4+}, etc.) metal cations or clusters, namely secondary building units (SBUs) (e.g., $Zn_4O(COO)_6$, $Cu_2(COO)_4$, $Cr_3O(H_2O)_3(COO)_6$, and $Zr_6O_4(OH)_{10}(H_2O)_6(COO)_6$) [10]. These inorganic units are connected by coordination bonds to organic linkers, which commonly contain carboxylate, phosphonate, pyridyl, and imidazolate, or other azolate functional groups [11].

Different synthesis methods have been developed and applied to synthesize MOFs to achieve the required properties in terms of shape, surface morphology, and others for various applications. The major synthesis methods of MOFs include hydrothermal (solvothermal) synthesis, mechanochemical synthesis, microwave-assisted synthesis, sonochemical synthesis, and electrochemical synthesis [12–15].

Solvothermal/hydrothermal synthesis is the most conventional method for the synthesis of these compounds. If water is used as a solvent, this method is termed as a hydrothermal method. This method generally consists of the self-assembly of soluble precursors, which contain the metal and the organic component. The key point of this method is the selection of the appropriate solvent because the solvent acts not only as a reaction medium but also as a structure-directing agent. Some examples of this synthesis include MOF-5 [16], MOF-74 [17], ZIF-8 [18], and Uio-66 [19]. The advantages of this technique are a large operating temperature range, relatively easy transposition, and high yield. Disadvantages include high energy consumption and a long reaction time [20].

Another commonly used method is mechanochemical synthesis, which is a green approach performed by grinding a mixture of the organic linker and metal salt in an agate mortar and pestle or in a ball mill and in the absence of solvent. Examples of this method include HKUST-1 [21], ZIF-4, and ZIF-8 [22].

In microwave-assisted synthesis, energy for the reaction is provided in the form of microwave radiation (MW). Advantages include a reduction in crystallization time and higher yields, on the other hand, disadvantages include difficulty in controlling the shape and size and the lack of a direct method for scale-up [20]. Examples include CoMOF-74 [23] and Cr-MIL-100 [24].

The ultrasonic/sonochemical method is another efficient synthesis of MOFs under ultrasonic irradiation. The impact of ultrasound on liquid and colloid systems is mainly caused by acoustic cavitation, which would be the production, growth, and implosion of bubbles within the solvent. The advantages of this method include homogeneous particle size and morphology in short periods of time and the formation of stable structures due to the high energy and high pressure created during this process [20]. Fe-MIL-53 [25] and MOF-177 [26] are some examples of MOF materials synthesized through this ultrasonic method.

In the electrochemical method, metal ions rather than metal salts are introduced in the anode, whereas an organic linker is placed at the cathode and the electrochemical cell is filled with conducting salt. This technique was reported by BASF in 2005. The advantages of this method are avoiding anions, such as nitrates, from metal salts, lower temperatures of reaction, and extremely quick synthesis [12]. Some examples of this synthesis are ZIF-8 and MIL-53 (**Table 1**) [27].

Synthetic method	Advantages	Disadvantages	Examples	Ref.
Hydro(solvo)thermal	1. Large operating temperature range (i.e., between 80 and 250°C) 2. Easy transposition 3. High yield	1. High energy consumption 2. Long reaction time 3. Expensive pressure-sealed metal vessels and heating ovens	MOF-5 ZIF-8	[16, 18]
Mechanochemical	1. Solvent-free synthetic method 2. No need for temperature and pressure	1. Difficult isolation single-crystals for X-ray diffraction studies 2. Formation of Secondary phases	HKUST-1 ZIF-4	[21, 22]
Microwave	1. Reduction in crystallization time 2. High yields 3. Possibility to control morphology, phase selectivity, and particle distribution 4. Simple method and easy to control the reaction parameters	1. Difficult to isolate large single crystals 2. The lack of a direct method for scale-up	CoMOF-74 Cr-MIL-100	[23, 24]
Ultrasonic	1. Homogeneous particle size and morphology 2. Short reaction time 3. Formation of stable structures 4. Suitable method for the preparation of nanosized MOFs	Destruction of crystallites hindering the formation of large single crystals for X-ray diffraction studies	Fe-MIL-53 MOF-177	[25, 26]
Electrochemical	1. Avoiding anions such as nitrates from metal salts 2. Low temperatures of reaction and extremely quick synthesis	Rarely reported method for synthesis of MOFs	ZIF-8 MIL-53	[27]

Table 1.
The advantages and disadvantages of the different methods.

3. Applications of MOFs

MOFs have been attracting great attention in the past several decades mainly because of their outstanding properties, including high specific surface area, ultra-high porosity, tunable surface chemistry, functional diversity, and crystallinity [9]. These properties make them good candidates for use in gas storage and separation [28, 29], catalysis, drug delivery, and sensing [30].

Common methods of gas storage require high-pressure tanks and multistage compressors, which are difficult and expensive. MOFs provide a simpler and safer storage method, owing to their high surface area, high porosity, and tunable pore structure. To date, numerous MOFs have been applied for gas storage. PCN-12 [31], MOF 210 [32], and NU-100 [33] are some examples for H_2 storage. MOFs also have great potential to perform gas separations as adsorbents. In 2014, Yang et al. reported a hydroxyl–functionalized Al-MOF (NOTT-300) with high selectivity for C_2H_4/C_2H_6 separation [34].

MOFs have shown great capability in heterogeneous catalysis due to their unique features. Numerous MOFs have been reported as catalysts in photocatalysis reactions, electrocatalysis reactions, and organic transformations, such as oxidation, epoxidation, and Knoevenagel condensation [35]. These compounds are commonly employed as heterogeneous catalysts in the form of pristine solids or post-synthetically modified ones. One of the earliest studies in the field of MOF catalysis was reported in 1994 by Fujita et al. They constructed a two-dimensional (2D) network of $Cd(4,4'\text{-}bpy)_2(NO_3)_2$ by reaction between $Cd(NO_3)_2$ and (4,4′-bpy), and used it as a heterogeneous catalyst for the cyanosilylation of aldehydes [36]. The first experimental evidence of photocatalytic activity on a MOF (MOF-5) was published in 2007 [37]. To date, many researches have been made in the area of MOF-based photocatalysts. For example, Leng et al. reported an indium-based MOF, USTC-8(In), which exhibits an excellent photocatalytic activity for H_2 production [38].

Researches on MOFs as drug carriers have been developed in biomedical fields due to their features, such as drug loading ability, biodegradability, and adaptable functionality. According to the enhanced stability, enormous porosity, and large pore volume of the MIL family, they are an attractive candidate for the storage and controlled release of biologically important molecules [39]. Serrey and Férey et al. demonstrated encapsulating drug molecules (Ibuprofen) in chromium carboxylate MOFs, MIL – 100 and MIL – 101, exhibiting drug storage capacities of 35 wt% and 140 wt%, respectively, and controlled drug release behavior of 5 to 6 days under physiological conditions [40].

In addition to the applications mentioned above, MOFs can be used as sensors in different fields. A sensor is a device that can receive, respond, and detect a signal. In MOF-based devices, the sensor responds to different external stimulations, such as stress, mass, optical, and environmental changes. MOFs, which can be excited by absorbing UV–visible light, are considered as photoluminescent. Both the metal and the linker increase luminescence. Lanthanides like Eu, Tb, Dy, Sm, Nd, Gd, Er, and Yb are employed as luminescent metal ions owing to their electronic transition from d- to f-orbitals. Naphthalene, anthracene, pyrene, perylene, and stilbene types of ligands are used for the synthesis of luminescent MOFs [15]. As an example, HKUST-1 has shown a color change from turquoise to green as a result of the exchange of Cu^{2+}-coordinated water with pyridine (**Table 2**) [44].

MOF	Metal node	Linker	Application	Ref.
UiO-66	$Zr_6O_4(OH)_4$	Benzene-1,4-dicarboxylic acid	Ring-opening reaction cyclization	[41, 42]
HKUST-1	Cu(II)	1,3,5-Benzene tricarboxylate	Esterification	
Ti-MOF-$Ru(tpy)_2$	Ti(IV)	Bis(4′-(4-carboxyphenyl)-terpyridine)Ru(II) complex	Photocatalytic H_2 production	[43]
PCN-12	Cu(II)	5,5′-Methylene-di-isophthalate	H_2 storage	[31]
Al-MOF (NOTT-300)	Al(III)	Biphenyl-3,3′,5,5′-tetracarboxylic acid	C_2H_4/C_2H_6 separation	[34]
MIL - 101	Cr(III)	1,4-Benzenedicarboxylate	Drug delivery	[40]
HKUST-1	Cu(II)	Benzene-1,3,5-tricarboxylate	sensor	[44]
Al-ATA-Ni MOF	Al(III)	2-Aminoterephthalate	Photocatalytic water splitting	[45]
Zr-MOF,	$Zr_6O_4(OH)_4$	Dibenzoate-substituted 2,2′-bipyridine dibenzoate	Oxidation	[46]
IRMOFs (MOF-5)	Zn_4O	Benzene-1,4-dicarboxylic acid	Alkylation	[47]

Table 2.
Different applications of MOF.

4. Different types of MOFs

MOFs can be divided into different types based on their metal nodes. Metal nodes may vary from the s-block, p-block, transition metals, or even rare earth metals. S-block main group elements, alkali and alkaline earth metals are rarely used because of their low stability and difficulties in the structural formation of MOFs. In spite of their limitations, they have some advantages, including naturally high abundance, less toxicity, and low density, which lead to different applications [48]. Most reported MOFs are based on transition metals that can divide into different groups.

4.1 Transition-metal-based MOFs

4.1.1 Zr-based MOFs

Zr-based MOFs are useful in different fields according to their extraordinary mechanical, thermal, chemical stability and high oxidation state of Zr(IV) compared with M(I), M(II), and M(III)-based MOFs (M stands for metal elements). Zr-MOFs are applied in catalysis, molecule adsorption and separation, drug delivery, and fluorescence sensing, and as porous carriers [49]. For instance, Feng et al. reported Zr-PCN-221(Fe) as the catalyst for the selective oxidation of cyclohexane with tert-butyl hydroperoxide (TBHP) as the oxidant [50]. Additionally, Zr-based MOFs can act as a semiconductor in photocatalytic reactions. As an example, Lillerud and coworkers reported the first Zr-based MOFs including UiO-66 and UiO-66(NH_2) for photocatalytic water splitting [51].

4.1.2 Cu-based MOFs

Copper (Cu) as an abundant and inexpensive metal with divalent copper ions having tetra- or hexa-coordinated structure can be used as another metal node in the synthesis of MOFs. Also, it has different oxidation states, such as Cu^0, Cu^I, Cu^{II}, and Cu^{III}, leading to a wide variety of redox reactions [52].

Park and coworkers reported the green synthesis of PNU-25 applied in the cycloaddition reaction of CO_2 and epoxide [53]. In another study, a semi-conductive copper organic MOF, that is, $\{[Cu^ICu^{II}_2\text{-}(DCTP)_2]NO_3{\cdot}1.5\ DMF\}_n$ with 2.1 eV bandgap was involved in UV/visible photocatalytic H_2 generation and oxidation of holes for photocatalytic degradation of methyl blue with use of visible light and without the aid of photosensitizer [54].

4.1.3 Fe-based MOFs

According to the Earth's abundance of iron, development in various applications of Fe-based MOFs in CO_2 reduction, photo-degradation, water splitting, and organic transformations occurred [55]. They are widely applied in photocatalytic reactions like H_2 production. For example, the photocatalytic visible light H_2 production of MIL-100(Fe) in presence of MEOH was reported by Wang et al. [56].

4.1.4 Co-based MOFs

Based on the inexpensiveness and availability of cobalt salts, cobalt-based MOFs have been widely used in different applications, such as oxygen and hydrogen evolution, catalysis, electrocatalysis, synthesis of nanomaterials, and more [57–59]. These structures can provide a transition porous framework and ultra-high surface area for photocatalytic H_2 production. For instance, Yang et al. reported Co-based zeolitic imidazolate MOF (ZIF-67) with RuN_3 as a photosensitizer for efficient photocatalytic H_2 production from H_2O [60]. In another study of Co-MOFs, Dirk Volkmer and coworkers reported the synthesis and the catalytic performance of MFU-1 MOF in the oxidation of cyclohexene using TBHP as an oxidant [61].

4.1.5 Ni-based MOFs

Ni-based MOFs are another useful group of metal–organic frameworks with properties such as high porosity, presence of coordinatively unsaturated Ni^{2+} sites and Lewis acid sites so they can be used in different fields. Miyake et al. first described the synthesis and application of MOFs with Ni-bipyridyl complex as the linker along with Ni as the nodes in ethylene oligomerization [62]. In another study, $[Ni(phen)\ (oba)]_n$- $0.5nH_2O$ 2D MOF has been studied as supporting material for metal sulfides as a visible light active photocatalyst for effective H_2 production [63].

4.1.6 Ti-based MOFs

4.1.6.1 The overall properties of titanium (Ti)

4.1.6.1.1 The chemistry of Ti

What makes titanium an attractive candidate to use as a metal node in several MOFs? Indeed, divalent transition metal cations, such as Zn^{2+}, Cu^{2+}, Co^{2+}, and Ni^{2+}, lead to thousands of extraordinary porous MOFs. Nevertheless, they are often unstable in the presence of water. To enhance the strength of cation–ligand bond and gain stability in MOFs, increasing the charge of a metal cation and its polarizing power charge over ionic radius are considered [64].

Among the elements in the periodic table, titanium or Ti has drawn considerable attention due to its fascinating intrinsic features. To mention about its abundance, it is the ninth most abundant element in the Earth's crust and is the second most abundant transition metal after iron. This element, located in the first row of the periodic table and above the zirconium, has a [Ar] $3d^2$ $4s^2$ electronic configuration. In comparison with Zr^{4+}, it has a stronger interaction with oxygen due to its smaller ionic radius. Its common oxidation states are +2, +3, and + 4. Ti^{+4} ion is a frequent oxidation state, which is the most stable in ambient conditions (air and water), while Ti^{3+} could be generated under reductive conditions but quickly oxidized back to Ti^{4+} in the presence of oxygen. The reasons for its tendency to hydrolyze into various titanium oxo clusters, or directly to generate TiO_2 precipitation are low electronegativity and high polarizability [65–67].

Due to the all features mentioned above, titanium is a highly desirable alternative to constructing stable and diverse MOFs. There are several Ti-MOFs that were published, including MIL-91 [68], MIL-125 [69], NTU-9 [70], MIL-101 [71], COK-69 [72], Ti-CAT-5 [73], MIL-167 [74], MIP-177 [75], MIL-100 [76], Ti-TBP [77], and ZSTUs [78], applied in different fields. However, Ti-MOFs have seldom been reported due to some reasons, which will be summarized into three main parts [79]:

i. Strong interaction between reactive titanium sources and ligand as a result of the high charge to radius (Z/r) value on Ti^{4+} causes poor association/dissociation of metal–ligand bond and formation of crystalline products. To achieve a crystalline structure, choosing suitable synthetic conditions, such as the organic solvents, ratio of the reacted solution mixture, pH, and temperature, should be considered.

ii. Another limitation is the extreme hydrolysis of reactive titanium sources

iii. Some of the titanium carboxylates show low symmetry or inappropriate connectivity that leads to hinder the formation of periodic networks with organic linkers [80].

4.1.6.1.2 History

MIL-22 was the first Ti-MOF, synthesized by Serre, Fèrey, and coworkers, using phosphorus ligand and TiO_2 under high temperature in 1999 (**Figure 1a**) [81]. This

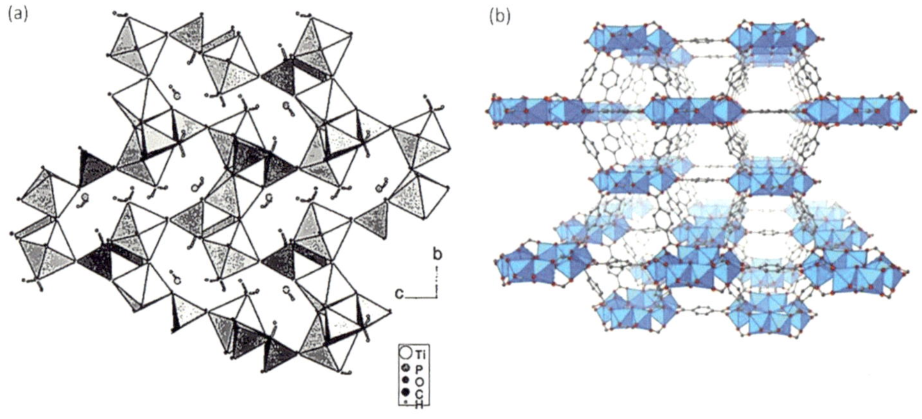

Figure 1.
(a) Structure of MIL-22, (b) structure of MIL-125.

compound showed limited porosity. For a quite long time, no highly porous Ti-MOF was reported. Subsequently, in 2009, MIL-125 was discovered by the same group (**Figure 1b**) [82]. The discovery of this MOF created a scientific boom in the synthesis of new Ti-based MOFs material. It consists of 12-connected $[Ti_8O_8(OH)_4(COO)_{12}]$ clusters and 1,4-benzenedicarboxylate (BDC) linker and demonstrates permanent porosity with a BET surface area of 1550 m^2g^{-1} [69]. Meanwhile, titanium oxo clusters, which can be seen as TiO_2 nanoclusters, affect the construction of novel Ti-based MOFs [83].

Up to now, different types of Ti-MOFs have been introduced. The development of a titanium-based MOFs timeline is presented in **Figure 2** [65].

4.1.6.1.3 Comparison of the Ti-MOF with other MOFs

In this regard, Ti-MOFs possess several advantages over other groups of MOFs, which are as follows:

i. Strong Ti–O bonding, which is the result of the small ionic radius of Ti^{4+}, leads to rigid MOF structures.

ii. The high coordination number of Ti^{4+} causes structural diversity in metal-oxo clusters.

iii. Ti-based MOFs with large surface area and high porosity can be effective photocatalysts in which Ti–O clusters act as isolated titanium oxide quantum dots, making them like the Ti-grafted zeolites and the semiconducting titanium oxide.

iv. Their ability to store charge effectively because of the reversible redox transition between Ti^{4+}/Ti^{3+} under photo-excitation leads to using them in photo-driven catalytic redox reactions.

v. Their photocatalytic properties can be developed and improved by functionalization of ligands and composition with other highly active species like metal nanoparticles and graphene.

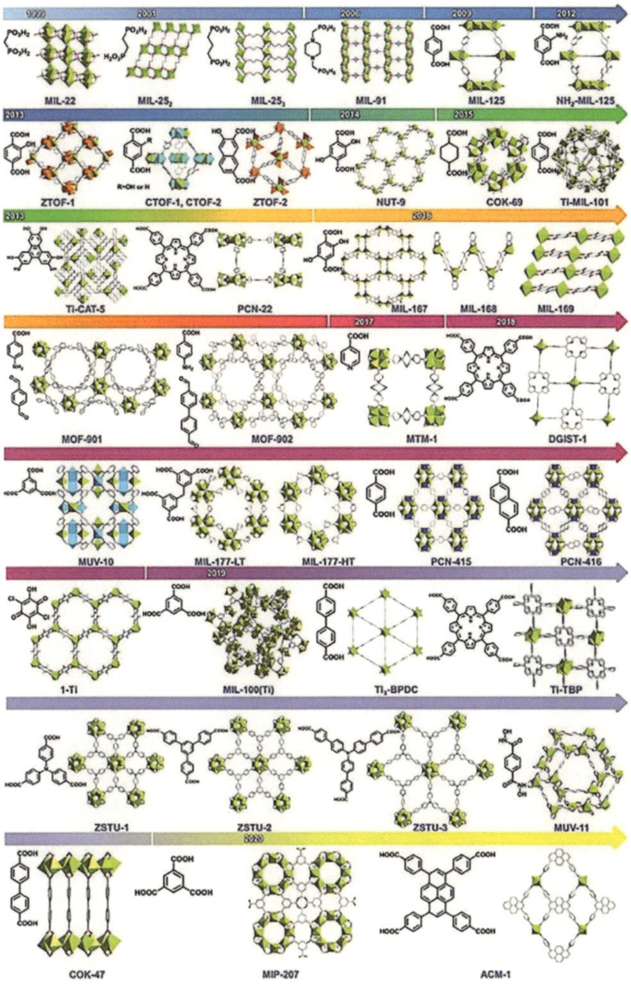

Figure 2.
The timeline for the development of crystal structures of titanium-based MOFs.

vi. They form rigid frameworks, which are the result of strong metal–ligand bonding.

vii. Their well-defined porosity enhances photocatalytic activities.

viii. Their low cost and low toxicity make them good candidates as catalysts [64, 84].

To realize the importance of Ti-MOFs among all families of MOFs, their properties should be discussed, most notably about two important ones such as porosity and optical properties [85].

Porosity is one of the most effective features in photocatalytic reactions accomplished by MOFs. Owing to this property, the transfer of photoelectric charge happens rapidly, thus the electron–hole recombination reduces. Moreover, porosity lets the active sites to be in contact with the substrate more effectively by reducing the diffusion resistance [86]. In addition, this property prepares large spaces for encapsulating different types of catalysts [87]. As a result, Ti-MOFs can be used in different fields of chemistry.

There are various strategies used for the synthesis of porous Ti-MOFs, which are good to mention:

i. Partial post-synthetic metal exchange; for instance, porous Ti^{III}-MOF-5 and Ti^{IV}-UiO-66 were synthesized by exchanging their metals with Zn^{II} and Zr^{IV} [88].

ii. Assembly of clusters; MTM-1, a porous Ti-MOF, is synthesized by a combination of $\{[Ti_6O_6][Oipr]_6\}^{6+}$ cluster and cross-connected Cu_2I_2 dimer through isononanol (INA) [85].

iii. The supramolecular template strategy [86].

iv. Isoreticular expansion, an effective strategy, is used to increase the pore size, surface area, and porosity by self-assembly of metal centers and organic ligands. In this strategy, long and geometrically equivalent organic ligands influence the porosity of MOF. As an example, MOF-902 has a greater porosity in comparison with MOF-901 because of its longer organic building blocks [89]. In spite of the fact that the longer the organic ligands are, the higher porosity occurs, and the stability of Ti-MOFs diminishes.

The optical property of MOFs is an important factor for photocatalysis. In general, photocatalytic reactions consist of three main steps: light absorption, charge separation and transfer, and redox reaction [90]. To enhance the light absorption intensity, modification of organic ligands is utilized. In the same way, it can affect MOF's bandgap. As an example, inserting the $-NH_2$ functional group on the ligand of MIL-125 yields a small bandgap [91].

To conclude, compared to other MOFs, Ti-MOFs are fascinating structures with various applications in different fields due to their high thermal and chemical stability, novel structural diversity, photocatalytic properties, low cost, and low toxicity (**Figure 3**) [85]. The geometry and length of organic building blocks and metal nodes can affect the chemical stability and functional properties of MOFs. For instance, the carboxylate ligand-based MOFs with low-valent ions, such as Zn^{2+}, Co^{2+}, and Mn^{2+}, show lower chemical stability. However, those with high-valent ions, such as Zr^{4+}, Ti^{4+}, and Cr^{3+}, usually have high chemical stability [65].

In comparison with other metal-based MOFs, for example, Zr-MOFs, Fe-MOFs, or Cu-MOFs, the titanium-based MOFs exhibit richer structural and topological diversity, due to the variable nuclearity of titanium cores.

Owing to the low-level control of titanium chemistry and highly unpredictable polycondensation of titanium cation, titanium-based MOF structures are less commonly reported compared with the Zr-MOFs, Fe-MOFs, or Cu-MOFs.

4.1.6.1.4 Stability

The stability of M^{4+}-MOFs makes them an appealing choice to concentrate on. Many factors are involved in the stability of MOFs, such as metal nodes, organic ligands, metal–ligand coordination, operating environment, hydrophobicity of the pore surface, and defects [92–94]. Thermodynamic stability of MOFs is affected by

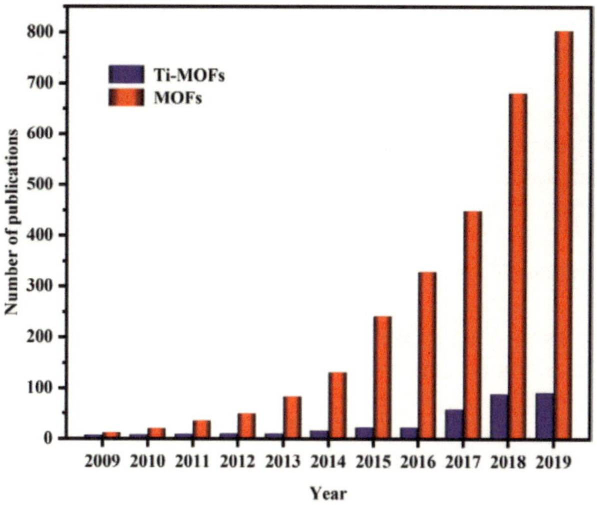

Figure 3.
The number of photocatalytic publications on MOFs and Ti-MOFs reported annually from 2009 to 2019.

metal–ligand coordination. Based on Pearson's hard/soft acid/base (HSAB) theory, strong bonds are formed between hard Lewis's acids and bases or soft Lewis's acids and bases. Carboxylate-based ligands are hard bases that can generate stable MOFs with group 4 metal cations such as Ti^{4+} [66]. Likewise, kinetic factors have an important role in the stability of MOFs. Strong connection between metal-oxo clusters and rigid ligands leads to dense and rigid frameworks, which increase the stability of MOFs. Hence, good stability in M^{4+}-based MOFs is expected because M^{4+} cations tend to have more ligands to balance their charges. Another factor that affects the stability of MOFs is the operating environment. For instance, the resistance of carboxylate-based ligands in acids and bases is completely different. They show excellent stability in acids, while their stability is quite weak in bases [95].

4.1.6.1.5 The coordination behavior of titanium ions

Generally, each metal ion prefers to coordinate with other components in a special way. For example, Cu^{2+} ions tend to assemble with four carboxylate groups to form $Cu(COO)_4$ paddle-wheel geometry [44] and zirconium ions construct the clusters like $Zr_6O_4(OH)_4(CO_2)_{12}$ as 6-C, 8-C, 10-C, or 12-C nodes [49]. With regard to titanium ions, the coordination with other components changes with reaction systems. The diversity in titanium-oxo clusters makes it hard to predict the topology of titanium-based MOFs. Ti^{4+} constructs octahedrally six-coordinated structures by O forming TiO_6 units that are bridged by μ_2-O or μ_3-O atoms. They lead to the creation of titanium-oxo clusters that vary based on their synthetic conditions. For instance, the cluster in MIL-125 or NH_2-MIL-125, $Ti_8O_8(OH)_4$ is completely different from $Ti_3(\mu_3$-O) cluster in its polymorph Ti-MIL-101 [69, 71].

4.1.6.1.6 Applications

Most of the investigations have demonstrated that Ti-MOFs have several applications due to their diversity of stable structures, including photocatalytic reactions and organic transformations such as H_2 productions [96], CO_2 conversion [97], H_2O_2

production [98], pollutant degradation [99], and many other organic reactions. As an example, in 2018, an efficient photocatalytic H_2 production, accomplished by a Ti-MOF known as MIL-125-$(SCH_3)_2$, was reported by Han et al. [100]. In 2018, Yamashita and coworkers reported the first example of a Ti-based MOF, MIL-125-NH_2 with NiO nanoparticles (NPs), used in photocatalytic H_2O_2 production [101]. The photocatalytic reduction of CO_2 with Co/NH_2-MIL-125(Ti) was reported by Fu et al. under visible light irradiation [102].

4.1.6.2 Titanium precursor for MOF synthesis

Titanium precursors used in MOFs synthesis include hydrous TiO_2, titanium chlorides, titanium alkoxides, titanium complexes, and titanium-oxo-clusters.

MIL-25 [103] and MIL-22 [81], two Ti natural phosphonates, were the first structures that were formed by hydrous TiO_2 as a precursor. Hydrous forms of TiO_2 are rarely used for the synthesis of MOFs due to their limitations, such as high reaction temperature (>200 C), long reaction time (3–4 d), and highly corrosive and toxic reaction media (40% HF aqueous solution) [104].

Titanium chlorides extremely hydrolyze and release hydrochloride fumes in the reaction medium [105]. This is the reason why they are seldom used in Ti-MOFs synthesis. $TiCl_3$ is an exception in the construction of di or trivalent Ti-MOFs. Whereas Ti^{4+} is not a suitable selection because of the imbalanced charge and ionic range. According to the reductive nature of Ti^{3+}, exposure to air inhibits the synthesis [106].

In terms of titanium alkoxides, $Ti(OR)_4$, hydrophobic character and hindrance of alkoxy groups cause them less hydrolyzed than $TiCl_4$. The construction of complexes of Ti cations with bidentate ligands, such as carboxylates, diketonates, catecholates, and phosphonates, leads to a stable structure [64].

An efficient way to control the reaction degree of MOF synthesis is the utilization of stable six-coordinated Ti compounds (Ti-oxo-clusters) [80]. Owing to the different sizes, geometry, and coordination numbers of Ti clusters, they have the potential to be chosen as building blocks of MOFs. The original cluster might be transformed during the reaction. For instance, in the synthesis of PCN-22 by $Ti_6O_6(iOPr)_6(abz)_6$ as Ti precursor, this cluster was completely transformed to Ti_7O_6-oxo SBU in the final MOFs [107]. In comparison with Ti alkoxides, Ti clusters are useful to adjust the reaction rates and obtain larger single crystals [106].

4.1.6.2.1 Synthesis methods

New synthetic methods of Ti-MOFs have been proposed to remove the restrictions. In this section, we have introduced some strategies to synthesize these structures, including direct synthesis, cooperative self-assembly strategy, and vapor-assisted crystallization method. Besides, the synthetic strategies of Ti-MOF composites and derived porous materials are discussed [85].

4.1.6.2.1.1 Direct synthesis

Most of the Ti-MOFs have been synthesized by direct methods, such as hydrothermal/solvothermal, microwave, and ultrasonic synthesis. The most conventional method for direct synthesis of Ti-MOFs is the hydrothermal/solvothermal technique, which includes the interaction between metal–salt and an organic linker in a suitable solvent(s) and autoclave conditions. Choosing the proper solvent and concentration of reactants are two important factors that affected the crystallinity and morphology of Ti-MOFs [106]. As an example, the morphology of NH_2-MIL-125 microcrystals could be changed from circular plate to octahedron by changing the concentration of the solution [108]. In 2015, Yaghi and coworkers reported the first three-dimensional (3D) extended Ti catecholate MOF (Ti-CAT-5) by solvothermal reaction of the Ti ion and hexatopic catecholate linker, 2,3,6,7,9,11-hexahydroxytriphenylene (H_6THO) at 180 C for 4 h [73].

Other direct methods for synthesis of Ti-MOFs are ultrasonic- and microwave-assisted techniques that have advantages including obtaining structures with different molecular sizes and shapes in short reaction time and low temperature [104]. Kim and coworkers have synthesized NH_2-MIL-125 by the microwave-assisted method in a shorter reaction time in comparison with the solvothermal technique. As an example of sonochemical synthesis, Han and coworkers have reported NH_2-MIL-125 in a uniform size distribution of around 300 nm [109].

4.1.6.2.1.2 Coordination-covalent combination method

In spite of the advantages of direct synthesis of Ti-MOFs, this methodology has some limitations like the formation of unpredicted structures and byproducts. The utilization of Ti-oxo clusters can approximately solve this problem. To obtain an ideal structure that does not disturb the Ti–O bonding in the clusters, combining the MOFs and COFs known as the coordination-covalent combination strategy was introduced [104]. The first example of this strategy was reported by Yaghi and coworkers. This reported synthesis involves *in situ* generation of an amine-functionalized titanium oxo cluster, $Ti_6O_6(OCH_3)_6(AB)_6$ (AB = 4-aminobenzoate), which was linked with benzene-1,4-dialdehyde using imine condensation reactions, typical of COFs. The crystal structure of MOF-901 with known structure and topology was achieved [110].

4.1.6.2.1.3 Post-synthetic cation exchange method

Another strategy to overcome the conventional methods' limitations is the cation exchange method. This technique can avoid the high reactivity and vigorous hydrolysis of Ti precursors and lead to desired pore and topography in Ti-MOFs [111]. The example for post Ti-exchange was first reported by Cohen et al. Ti ion exchange occurred in Zr-MOF, UiO-66, became a better option for Ti-MOF [112].

4.1.6.2.1.4 Vapor-assisted crystallization method

To overcome the limitations of microporous structures in catalytic applications of Ti-MOFs by creating mesopores in MOFs structures, the vapor-assisted crystallization method (VAC) was introduced by Hicks et al. [113].

4.1.6.2.1.5 Synthesis of Ti-MOF composites

According to the importance of MOFs supported metal nanoparticle compounds in catalysis, highly dispersed nanoparticle-doped Ti-MOFs were prepared. Not only noble metal nanoparticles, such as silver (Ag), gold(Au), platinum (Pt), palladium (Pd), and non-noble metal nanoparticles, such as nickel (Ni) have been successfully embedded into the pores of Ti-MOFs [114, 115] but also quantum dots, such as graphene-like MoS_2 sheets, Ag_2S, CdS, and CuS have been deposited onto MIL-125 and a series of Ti MOFs based composites [116].

4.1.6.2.2 Different types of Ti-MOFs

Ti-MOFs can be categorized based on their structures (one-dimensional (1D) chain structures, two-dimensional (2D) layer structures, and three-dimensional (3D) network structures) and ligands in their secondary building units (phosphonates, carboxylates, salicylates, and catecholates).

4.1.6.2.2.1 Types of Ti-MOFs based on their structures

1D and 2D Ti-MOF structures are rarely reported. For instance, MIL-168 has 1D zigzag chains running parallel to each other (**Figure 4a**) [74]. To form an ordered 1D zigzag, a single bidentate ligand occupies two cis positions to create an octahedral environment around the metal ion [117].

Generally, 2D network structures can be divided into two types: "tilt penetration" and "parallel penetration" [118]. In two-dimensional MIL-169 ($Ti_2(DOBDC)_4O(H_2O)_2$ dimer, $DOBDC_4$ = 2,5-dioxido-1,4-benzenedicarboxylate), the Ti ion is chelated by two DOBDC ligands through the six-membered ring (**Figure 4b**) [74].

In terms of 3D MOFs, they possess different complicated structures. For example, Jiang et al. reported the synthesis and structure of a 3D chiral structure, a 3-connected $SrSi_2$ net with (10,3) topology, which exhibited a novel doubly interwoven $SrSi_2$ topology (**Figure 4c**) [119].

4.1.6.2.2.2 Types of Ti-MOFs based on ligands in secondary building units

It is noteworthy to discuss about the structures of Ti-MOFs based on different kinds of ligands due to their obvious directing effects toward the finally formed Ti-oxo SBUs.

4.1.6.2.2.2.1 Carboxylate-based Ti-MOFs

In general, Ti-MOFs are mostly constructed from carboxylate ligands. Ti cations with polydentate carboxylate acids have a tendency to form large Ti-oxo-carboxylate clusters, leading to a wide variety of Ti-MOFs' structures. Ti-oxo-carboxylate clusters with the general formula $Ti_nO_m(OR)_x(OOCR')_y$ can be synthesized by a solution of a

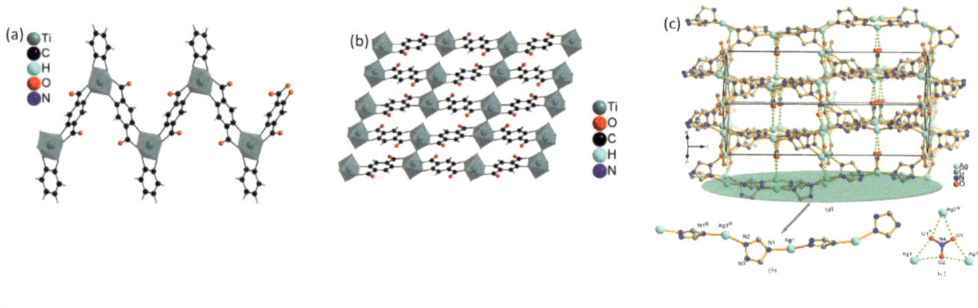

Figure 4.
(a) Crystal structure of MIL-168, a 1D coordination chain. (b) Crystal structure of MIL-169, 2D coordination network. (c) a view of the structure of poly[bis(μ_3–3,5-di-ethyl-1,2,4-triazolato-κ^3N^1:N^2:N^4)trisilver nitrate] along the [110] direction, showing the three-dimensional framework.

metal cation precursor (very often $Ti(OR)_4$) with carboxylate ligand and/or a controlled amount of water. It is important to control different factors such as nature and the amount of the metal cation precursor, the ligand (R$'$COOH), the solvent, and the reaction conditions (time, pressure, presence of water or not, inert conditions or not, etc.). In 2015, Bueken et al. reported a new Ti-MOF based namely COK-69 by reaction of trans-1,4-cyclo-hexanedicarboxylate (cdc) linkers and an unprecedented $[Ti^{IV}_3$ -(μ_3-O) $(O)_2(COO)_6]$ cluster. In COK-69, three neighboring Ti atoms joined by a μ_3-O^{2-} ion and six carboxylates give the formation of the Ti_3O_3 cluster. Subsequently, the porous framework is formed by a connection between the cluster and six cdc^{2-} linkers [120].

4.1.6.2.2.2.2 Phosphonate-based Ti-MOFs

A series of different phosphonate-based Ti-MOFs, typically MIL-22 [81], MIL-25_n (n = 2, 3) [103], and MIL-91(Ti) [68], has been introduced by Férey et al. For example, the monoclinic MIL-22 is synthesized by corner-shared trimers of Ti-oxo octahedrals and PO_4 from diphosphonic linkers, forming a 2D array of microporous channels.

4.1.6.2.2.2.3 Salicylate-based Ti-MOFs

As an example of using salicylate organic linkers in the structure of MOF, in 2013 Hong et al. introduced Ti-salicylate MOF, namely ZTOF-1. Under solvothermal conditions, the reaction of zinc and titanium precursors with asymmetrical dicarboxylate ligand, 2-hydroxyterephthalic acid (H_3obdc) leads to orange crystals of ZTOF-1. In this structure, Ti coordinates with the "salicylate" end of the ligand, while Zn coordinates with the carboxylate end [121].

4.1.6.2.2.2.4 Catecholate-based Ti-MOFs

Up to now only one Ti-MOF based on polycatecholate ligand known as Ti-CAT-5 was reported, in spite of the richness of Ti-catecholate clusters and the potential of these structures to apply in different fields. Ti-CAT-5 was constructed by the connection of isolated Ti-centers and the tritopic catechol derivative ligand, H_6THO [73].

5. Conclusion

Metal–organic frameworks, composed of metal ions or well-defined secondary building units and organic linkers, are utilized in different fields, such as gas storage and separation, liquid separation and purification, electrochemical energy storage, catalysis, and sensing, due to their unique properties. According to the features of titanium, such as different oxidation states, high coordination number of Ti^{4+}, and the ability to form strong bonds with oxygen and different ligands, it is used as a metal node in MOF design. Titanium-based MOFs, as a branch of metal–organic frameworks, have recently attracted the researchers' attention because of their special properties, including high thermal and chemical stability, novel structural diversity, photocatalytic properties, low cost, and low toxicity. In this regard, the characteristics, types of Ti-MOFs, methods of synthesis, applications, and comparison of Ti-MOFs with other MOFs are discussed in this chapter. In comparison with other metal-based MOFs, for example, Zr-MOFs, Fe-MOFs, or Cu-MOFs, the titanium-based MOFs exhibit richer structural and topological diversity, high thermal and chemical stability, and high photocatalytic activity.

Conflict of interest

The authors declare no conflict of interest.

Author details

Behnaz Delaram, Faezeh Mokari, Mostafa Gholizadeh and Yahya Absalan*
Faculty of Science, Department of Chemistry, Ferdowsi University of Mashhad, Mashhad, Islamic Republic of Iran

*Address all correspondence to: y.absalan@ferdowsi.um.ac.ir
and yahyaabsalan2014@gmail.com

References

[1] Corma A, García H, Llabrés FX, Xamena I. Engineering metal organic frameworks for heterogeneous catalysis. Chemical Reviews. 2010;**110**:4606-4655. DOI: 10.1021/CR9003924/ASSET/CR9003924.FP.PNG_V03

[2] Tomic EA. Thermal stability of coordination polymers. Journal of Applied Polymer Science. 1965;**9**:3745-3752. DOI: 10.1002/APP.1965.070091121

[3] Sowerby DB, Audrieth LF. Inorganic polymerization reactions III. Coordination polymerization. Journal of Chemical Education. 1960;**37**:134-137. DOI: 10.1021/ED037P134

[4] Berlin AA, Matveeva NG. Polymeric chelate compounds. Russian Chemical Reviews. 1960;**29**:119-128. DOI: 10.1070/rc1960v029n03abeh001223

[5] Hoskins BF, Robson R. Design and construction of a new class of scaffolding-like materials comprising infinite Polymeric frameworks of 3D-linked molecular rods. A reappraisal of the Zn(CN)2 and Cd(CN)2 structures and the synthesis and structure of the diamond-related frameworks [N(CH3)4][CuIZnII(CN)4] and CuI[4,4′,4″,4‴-tetracyanotetraphenylmethane] BF4·xC6H5NO2. Journal of the American Chemical Society. 1990;**112**:1546-1554. DOI: 10.1021/JA00160A038/SUPPL_FILE/JA00160A038_SI_001.PDF

[6] Batten SR, Hoskins BF, Robson R. Two interpenetrating 3D networks which generate spacious sealed-off compartments enclosing of the order of 20 solvent molecules in the structures of Zn(CN)(NO3)(tpt)2/3•solv (tpt = 2,4,6-tri(4-pyridyl)-1,3,5-triazine, solv = ~3/4C2H2Cl4•3/4CH3OH or ~3/2CHCl3•1/3CH3OH). Journal of the American Chemical Society. 1995;**117**:5385-5386. DOI: 10.1021/JA00124A032/SUPPL_FILE/JA5385.PDF

[7] Yaghi OM, Li H. Hydrothermal synthesis of a metal-organic framework containing large rectangular channels. Journal of the American Chemical Society. 1995;**117**:10401-10402. DOI: 10.1021/JA00146A033/SUPPL_FILE/JA10401.PDF

[8] Li H, Eddaoudi M, O'Keeffe M, Yaghi OM. Design and synthesis of an exceptionally stable and highly porous metal-organic framework. Nature. 1999;**4026759**(402):276-279. DOI: 10.1038/46248

[9] Kumar S, Jain S, Nehra M, Dilbaghi N, Marrazza G, Kim KH. Green synthesis of metal–organic frameworks: A state-of-the-art review of potential environmental and medical applications. Coordination Chemistry Reviews. 2020;**420**:213407. DOI: 10.1016/J.CCR.2020.213407

[10] Yuan S, Qin JS, Lollar CT, Zhou HC. Stable metal-organic frameworks with group 4 metals: Current status and trends. ACS Central Science. 2018;**4**:440-450. DOI: 10.1021/ACSCENTSCI.8B00073/ASSET/IMAGES/LARGE/OC-2018-000738_0005.JPEG

[11] Howarth AJ, Liu Y, Li P, Li Z, Wang TC, Hupp JT, et al. Chemical, thermal and mechanical stabilities of metal–organic frameworks. Nature Reviews Materials. 2016;**1**(3):1-15. DOI: 10.1038/natrevmats.2015.18

[12] Lee YR, Kim J, Ahn WS. Synthesis of metal-organic frameworks: A mini review. Korean Journal of Chemical

Engineering. 2013;**309**(30):1667-1680. DOI: 10.1007/S11814-013-0140-6

[13] Gangu KK, Maddila S, Mukkamala SB, Jonnalagadda SB. A review on contemporary metal–organic framework materials. Inorganica Chimica Acta. 2016;**446**:61-74. DOI: 10.1016/J.ICA.2016.02.062

[14] Butova VV, Soldatov MA, Guda AA, Lomachenko KA, Lamberti C. Metal-organic frameworks: Structure, properties, methods of synthesis and characterization. Russian Chemical Reviews. 2016;**85**:280-307. DOI: 10.1070/rcr4554

[15] Soni S, Bajpai PK, Arora C. A review on metal-organic framework: Synthesis, properties and application. Characterization and Application of Nanomaterials. 2018;**2**(2):87-106. DOI: 10.24294/CAN.V2I2.551

[16] McKinstry C, Cathcart RJ, Cussen EJ, Fletcher AJ, Patwardhan SV, Sefcik J. Scalable continuous solvothermal synthesis of metal organic framework (MOF-5) crystals. Chemical Engineering Journal. 2016;**285**:718-725. DOI: 10.1016/J.CEJ.2015.10.023

[17] Bhattacharjee S, Choi JS, Yang ST, Choi SB, Kim J, Ann WS. Solvothermal synthesis of Fe-MOF-74 and its catalytic properties in phenol hydroxylation. Journal of Nanoscience and Nanotechnology. 2010;**10**:135-141. DOI: 10.1166/JNN.2010.1493

[18] R. Banerjee, A. Phan, B. Wang, C. Knobler, H. Furukawa, M. O'Keeffe, O.M. Yaghi, High-throughput synthesis of zeolitic imidazolate frameworks and application to CO2 capture, Science (80-.). 319 (2008) 939-943. DOI: 10.1126/SCIENCE.1152516/SUPPL_FILE/BANERJEE.SOM.PDF

[19] Abid HR, Tian H, Ang HM, Tade MO, Buckley CE, Wang S. Nanosize Zr-metal organic framework (UiO-66) for hydrogen and carbon dioxide storage. Chemical Engineering Journal. 2012;**187**:415-420. DOI: 10.1016/J.CEJ.2012.01.104

[20] Silva P, Vilela SMF, Tomé JPC, Almeida Paz FA. Multifunctional metal–organic frameworks: From academia to industrial applications. Chemical Society Reviews. 2015;**44**:6774-6803. DOI: 10.1039/C5CS00307E

[21] Klimakow M, Klobes P, Thünemann AF, Rademann K, Emmerling F. Mechanochemical synthesis of metal-organic frameworks: A fast and facile approach toward quantitative yields and high specific surface areas. Chemistry of Materials. 2010;**22**:5216-5221. DOI: 10.1021/CM1012119/SUPPL_FILE/CM1012119_SI_001.PDF

[22] Beldon PJ, Fábián L, Stein RS, Thirumurugan A, Cheetham AK, Friščić T. Rapid room-temperature synthesis of Zeolitic Imidazolate frameworks by using Mechanochemistry. Angewandte Chemie International Edition. 2010;**49**:9640-9643. DOI: 10.1002/ANIE.201005547

[23] Cho HY, Yang DA, Kim J, Jeong SY, Ahn WS. CO2 adsorption and catalytic application of Co-MOF-74 synthesized by microwave heating. Catalysis Today. 2012;**185**:35-40. DOI: 10.1016/J.CATTOD.2011.08.019

[24] Sung HJ, Lee JH, Chang JS. Microwave synthesis of a Nanoporous hybrid material, chromium Trimesate. Bulletin of the Korean Chemical Society. 2005;**26**:880-881. DOI: 10.5012/BKCS.2005.26.6.880

[25] Haque E, Khan NA, Park HJ, Jhung SH. Synthesis of a metal–organic

framework material, iron terephthalate, by ultrasound, microwave, and conventional electric heating: A kinetic study. Chemistry–A European Journal. 2010;**16**:1046-1052. DOI: 10.1002/CHEM.200902382

[26] Jung DW, Yang DA, Kim J, Kim J, Ahn WS. Facile synthesis of MOF-177 by a sonochemical method using 1-methyl-2-pyrrolidinone as a solvent. Dalton Transactions. 2010;**39**:2883-2887. DOI: 10.1039/B925088C

[27] Martinez Joaristi A, Juan-Alcañiz J, Serra-Crespo P, Kapteijn F, Gascon J. Electrochemical synthesis of some archetypical Zn 2+, Cu 2+, and Al 3+ metal organic frameworks. Crystal Growth & Design. 2012;**12**:3489-3498. DOI: 10.1021/CG300552W/SUPPL_FILE/CG300552W_SI_001.PDF

[28] Li H, Li L, Lin R-B, Zhou W, Zhang Z, Xiang S, et al. Porous metal-organic frameworks for gas storage and separation: Status and challenges. EnergyChem. 2019;**1**:100006. DOI: 10.1016/J.ENCHEM.2019.100006

[29] Ma S, Zhou HC. Gas storage in porous metal–organic frameworks for clean energy applications. Chemical Communications. 2010;**46**:44-53. DOI: 10.1039/B916295J

[30] He X. Metal-Organic Framework (MOF)-Based Materials: Aerosol Synthesis and Photocatalytic Applications. Theses Dissertation. 2019. pp. 3-9. DOI: 10.25772/63GS-6W17

[31] Wang X-S, Ma S, Forster PM, Yuan D, Eckert J, López JJ, et al. Enhancing H2 uptake by "close-packing" alignment of open copper sites in metal–organic frameworks. Angewandte Chemie. 2008;**120**:7373-7376. DOI: 10.1002/ANGE.200802087

[32] Furukawa H, Ko N, Go YB, Aratani N, Choi SB, Choi E, et al. Ultrahigh porosity in metal-organic frameworks. Science (80-.). 2010;**329**:424-428. DOI: 10.1126/SCIENCE.1192160/SUPPL_FILE/FURUKAWA-SOM.PDF

[33] Farha OK, Yazaydin AÖ, Eryazici I, Malliakas CD, Hauser BG, Kanatzidis MG, et al. De novo synthesis of a metal–organic framework material featuring ultrahigh surface area and gas storage capacities. Nature Chemistry. 2010;**211**(2):944-948. DOI: 10.1038/nchem.834

[34] Yang S, Ramirez-Cuesta AJ, Newby R, Garcia-Sakai V, Manuel P, Callear SK, et al. Supramolecular binding and separation of hydrocarbons within a functionalized porous metal–organic framework. Nature Chemistry. 2014;**72**(7):121-129. DOI: 10.1038/nchem.2114

[35] Jiao L, Wang Y, Jiang HL, Xu Q. Metal–organic frameworks as platforms for catalytic applications. Advanced Materials. 2018;**30**:1703663. DOI: 10.1002/ADMA.201703663

[36] Fujita M, Washizu S, Ogura K, Kwon YJ. Preparation, Clathration ability, and catalysis of a two-Dimensional Square network material composed of cadmium(II) and 4, 4′-Bipyridine. Journal of the American Chemical Society. 1994;**116**:1151-1152. DOI: 10.1021/JA00082A055/SUPPL_FILE/JA1151.PDF

[37] Alvaro M, Carbonell E, Ferrer B, Llabrés FX, Xamena I, Garcia H. Semiconductor behavior of a metal-organic framework (MOF). Chemistry–A European Journal. 2007;**13**:5106-5112. DOI: 10.1002/CHEM.200601003

[38] Leng F, Liu H, Ding M, Lin QP, Jiang HL. Boosting photocatalytic

hydrogen production of Porphyrinic MOFs: The metal location in Metalloporphyrin matters. ACS Catalysis. 2018;**8**:4583-4590. DOI: 10.1021/ACSCATAL.8B00764/SUPPL_FILE/CS8B00764_SI_006.CIF

[39] Keskin S, Kizilel S. Biomedical applications of metal organic frameworks. Industrial and Engineering Chemistry Research. 2011;**50**:1799-1812. DOI: 10.1021/IE101312K

[40] Horcajada P, Serre C, Vallet-Regí M, Sebban M, Taulelle F, Férey G. Metal–organic frameworks as efficient materials for drug delivery. Angewandte Chemie International Edition. 2006;**45**:5974-5978. DOI: 10.1002/ANIE.200601878

[41] Cai G, Jiang H-L. A modulator-induced defect-formation strategy to hierarchically porous metal–organic frameworks with high stability. Angewandte Chemie International Edition. 2017;**129**:578-582. DOI: 10.1002/ANGE.201610914

[42] Vermoortele F, Bueken B, Le Bars G, Van De Voorde B, Vandichel M, Houthoofd K, et al. Synthesis modulation as a tool to increase the catalytic activity of metal-organic frameworks: The unique case of UiO-66(Zr). Journal of the American Chemical Society. 2013;**135**:11465-11468. DOI: 10.1021/JA405078U/SUPPL_FILE/JA405078U_SI_001.PDF

[43] Sun CY, Liu SX, Liang DD, Shao KZ, Ren YH, Su ZM. Highly stable crystalline catalysts based on a microporous metal-organic framework and polyoxometalates. Journal of the American Chemical Society. 2009;**131**:1883-1888. DOI: 10.1021/JA807357R/SUPPL_FILE/JA807357R_SI_007.CIF

[44] Chui SSY, Lo SMF, Charmant JPH, Orpen AG, Williams ID. A chemically functionalizable nanoporous material [Cu3(TMA)2 (H2O)3](n). Science (80-.). 1999;**283**:1148-1150. DOI: 10.1126/SCIENCE.283.5405.1148/SUPPL_FILE/986116S4_THUMB.GIF

[45] An Y, Liu Y, An P, Dong J, Xu B, Dai Y, et al. NiII coordination to an Al-based metal–organic framework made from 2-Aminoterephthalate for photocatalytic overall water splitting. Angewandte Chemie. 2017;**129**:3082-3086. DOI: 10.1002/ANGE.201612423

[46] Wang C, Wang JL, Lin W. Elucidating molecular iridium water oxidation catalysts using metal-organic frameworks: A comprehensive structural, catalytic, spectroscopic, and kinetic study. Journal of the American Chemical Society. 2012;**134**:19895-19908. DOI: 10.1021/JA310074J/SUPPL_FILE/JA310074J_SI_001.PDF

[47] Ravon U, Domine ME, Gaudillère C, Desmartin-Chomel A, Farrusseng D. MOFs as acid catalysts with shape selectivity properties. New Journal of Chemistry. 2008;**32**:937-940. DOI: 10.1039/B803953B

[48] Alnaqbi MA, Alzamly A, Ahmed SH, Bakiro M, Kegere J, Nguyen HL. Chemistry and applications of s-block metal–organic frameworks. Journal of Materials Chemistry A. 2021;**9**:3828-3854. DOI: 10.1039/D0TA09678D

[49] Bai Y, Dou Y, Xie LH, Rutledge W, Li JR, Zhou HC. Zr-based metal–organic frameworks: Design, synthesis, structure, and applications. Chemical Society Reviews. 2016;**45**:2327-2367. DOI: 10.1039/C5CS00837A

[50] Feng D, Jiang HL, Chen YP, Gu ZY, Wei Z, Zhou HC. Metal-organic frameworks based on previously unknown Zr8/Hf 8 cubic clusters. Inorganic Chemistry.

2013;**52**:12661-12667. DOI: 10.1021/IC4018536/SUPPL_FILE/IC4018536_SI_005.PDF

[51] Cavka JH, Jakobsen S, Olsbye U, Guillou N, Lamberti C, Bordiga S, et al. A new zirconium inorganic building brick forming metal organic frameworks with exceptional stability. Journal of the American Chemical Society. 2008;**130**:13850-13851. DOI: 10.1021/JA8057953/SUPPL_FILE/JA8057953_SI_001.PDF

[52] Janczarek M, Kowalska E. On the origin of enhanced photocatalytic activity of copper-modified Titania in the oxidative reaction systems. Catal. 2017;7:317. DOI: 10.3390/CATAL7110317

[53] Kurisingal JF, Rachuri Y, Gu Y, Chitumalla RK, Vuppala S, Jang J, et al. Facile green synthesis of new copper-based metal-organic frameworks: Experimental and theoretical study of the CO2Fixation reaction. ACS Sustainable Chemistry & Engineering. 2020;**8**:10822-10832. DOI: 10.1021/ACSSUSCHEMENG.0C02749/SUPPL_FILE/SC0C02749_SI_001.PDF

[54] Wu ZL, Wang CH, Zhao B, Dong J, Lu F, Wang WH, et al. A semi-conductive copper–organic framework with two types of photocatalytic activity. Angewandte Chemie International Edition. 2016;**55**:4938-4942. DOI: 10.1002/ANIE.201508325

[55] Tasleem S, Tahir M, Khalifa WA. Current trends in structural development and modification strategies for metal-organic frameworks (MOFs) towards photocatalytic H2 production: A review. International Journal of Hydrogen Energy. 2021;**46**:14148-14189. DOI: 10.1016/J.IJHYDENE.2021.01.162

[56] Wang D, Song Y, Cai J, Wu L, Li Z. Effective photo-reduction to deposit Pt nanoparticles on MIL-100(Fe) for visible-light-induced hydrogen evolution. New Journal of Chemistry. 2016;**40**:9170-9175. DOI: 10.1039/C6NJ01989G

[57] Cai X, Peng F, Luo X, Ye X, Zhou J, Lang X, et al. Understanding the evolution of cobalt-based metal-organic frameworks in Electrocatalysis for the oxygen evolution reaction. ChemSusChem. 2021;**14**:3163-3173. DOI: 10.1002/CSSC.202100851

[58] Han W, Huang X, Lu G, Tang Z. Research progresses in the preparation of Co-based catalyst derived from Co-MOFs and application in the catalytic oxidation reaction. Catalysis Surveys from Asia. 2018;**232**(23):64-89. DOI: 10.1007/S10563-018-9258-1

[59] Zhang W, Huang R, Song L, Shi X. Cobalt-based metal–organic frameworks for the photocatalytic reduction of carbon dioxide. Nanoscale. 2021;**13**:9075-9090. DOI: 10.1039/D1NR00617G

[60] Yang S, Pattengale B, Kovrigin EL, Huang J. Photoactive Zeolitic Imidazolate framework as intrinsic heterogeneous catalysts for light-driven hydrogen generation. ACS Energy Letters. 2017;**2**:75-80. DOI: 10.1021/ACSENERGYLETT.6B00540/SUPPL_FILE/NZ6B00540_SI_001.PDF

[61] Tonigold M, Tu Y, Bredenkötter B, Rieger B, Bahnmüller S, Hitzbleck J, et al. Heterogeneous catalytic oxidation by MFU-1: A cobalt(II)-containing metal–organic framework. Angewandte Chemie International Edition. 2009;**48**:7546-7550. DOI: 10.1002/ANIE.200901241

[62] Kyogoku K, Yamada C, Suzuki Y, Nishiyama S, Fukumoto K, Yamamoto H, et al. Syntheses of metal-organic framework compounds containing Ni-bipyridyl complex for oligomerization of ethylene. Journal of the Japan

Petroleum Institute. 2010;**53**:308-312. DOI: 10.1627/JPI.53.308

[63] Ran J, Qu J, Zhang H, Wen T, Wang H, Chen S, et al. 2D metal organic framework Nanosheet: A universal platform promoting highly efficient visible-light-induced hydrogen production. Advanced Energy Materials. 2019;**9**:1803402. DOI: 10.1002/AENM.201803402

[64] Assi H, Mouchaham G, Steunou N, Devic T, Serre C. Titanium coordination compounds: From discrete metal complexes to metal–organic frameworks. Chemical Society Reviews. 2017;**46**:3431-3452. DOI: 10.1039/C7CS00001D

[65] Li L, Wang X, Liu T, Ye J. Titanium-based MOF materials: From crystal engineering to Photocatalysis. Small Methods. 2020;**4**:2000486. DOI: 10.1002/smtd.202000486

[66] Scarpi-Luttenauer M, Mobian P, Barloy L. Synthesis, structure and functions of discrete titanium-based multinuclear architectures. Coordination Chemistry Reviews. 2022;**459**:214439. DOI: 10.1016/J.CCR.2022.214439

[67] Buettner KM, Valentine AM. Bioinorganic chemistry of titanium. Chemical Reviews. 2012;**112**:1863-1881. DOI: 10.1021/cr1002886

[68] Serre C, Groves JA, Lightfoot P, Slawin AMZ, Wright PA, Stock N, et al. Synthesis, structure and properties of related microporous N,N′-piperazinebismethylenephosphonates of aluminum and titanium. Chemistry of Materials. 2006;**18**:1451-1457. DOI: 10.1021/cm052149l

[69] Dan-Hardi M, Serre C, Frot T, Rozes L, Maurin G, Sanchez C, et al. A new photoactive crystalline highly porous titanium(IV) dicarboxylate. Journal of the American Chemical Society. 2009;**131**:10857-10859. DOI: 10.1021/JA903726M/SUPPL_FILE/JA903726M_SI_002.CIF

[70] Gao J, Miao J, Li PZ, Teng WY, Yang L, Zhao Y, et al. A p-type Ti(IV)-based metal–organic framework with visible-light photo-response. Chemical Communications. 2014;**50**:3786-3788. DOI: 10.1039/C3CC49440C

[71] Mason JA, Darago LE, Lukens WW, Long JR. Synthesis and O2 reactivity of a titanium(III) metal-organic framework. Inorganic Chemistry. 2015;**54**:10096-10104. DOI: 10.1021/ACS.INORGCHEM.5B02046/SUPPL_FILE/IC5B02046_SI_001.PDF

[72] Bueken B, Vermoortele F, Vanpoucke DEP, Reinsch H, Tsou CC, Valvekens P, et al. A flexible photoactive titanium metal–organic framework based on a [TiIV3(μ3-O)(O)2(COO)6] cluster. Angewandte Chemie International Edition. 2015;**54**:13912-13917. DOI: 10.1002/ANIE.201505512

[73] Nguyen NTT, Furukawa H, Gándara F, Trickett CA, Jeong HM, Cordova KE, et al. Three-dimensional metal-Catecholate frameworks and their ultrahigh proton conductivity. Journal of the American Chemical Society. 2015;**137**:15394-15397. DOI: 10.1021/JACS.5B10999/SUPPL_FILE/JA5B10999_SI_004.CIF

[74] Assi H, Pardo Pérez LC, Mouchaham G, Ragon F, Nasalevich M, Guillou N, et al. Investigating the case of titanium(IV) Carboxyphenolate photoactive coordination polymers. Inorganic Chemistry. 2016;**55**:7192-7199. DOI: 10.1021/ACS.INORGCHEM.6B01060/SUPPL_FILE/IC6B01060_SI_004.CIF

[75] Wang S, Wang H, Ding WX. Pyroptosis, a novel player for alcoholic

hepatitis? Hepatology. 2018;**67**:1660-1662. DOI: 10.1002/HEP.29725

[76] Castells-Gil J, Padial NM, Almora-Barrios N, Da Silva I, Mateo D, Albero J, et al. De novo synthesis of mesoporous photoactive titanium(IV)–organic frameworks with MIL-100 topology. Chemical Science. 2019;**10**:4313-4321. DOI: 10.1039/C8SC05218B

[77] Lan G, Ni K, Veroneau SS, Feng X, Nash GT, Luo T, et al. Titanium-based nanoscale metal-organic framework for type i photodynamic therapy. Journal of the American Chemical Society. 2019;**141**:4204-4208. DOI: 10.1021/JACS.8B13804/SUPPL_FILE/JA8B13804_SI_002.PDF

[78] Li C, Xu H, Gao J, Du W, Shangguan L, Zhang X, et al. Tunable titanium metal–organic frameworks with infinite 1D Ti–O rods for efficient visible-light-driven photocatalytic H2 evolution. Journal of Materials Chemistry A. 2019;**7**:11928-11933. DOI: 10.1039/C9TA01942A

[79] Zou L, Feng D, Liu TF, Chen YP, Yuan S, Wang K, et al. A versatile synthetic route for the preparation of titanium metal-organic frameworks. Chemical Science. 2016;**7**:1063-1069. DOI: 10.1039/c5sc03620h

[80] Rozes L, Sanchez C. Titanium oxo-clusters: Precursors for a Lego-like construction of nanostructured hybrid materials. Chemical Society Reviews. 2011;**40**:1006-1030. DOI: 10.1039/C0CS00137F

[81] Serre C, Férey G. Hybrid open frameworks. 8. Hydrothermal synthesis, crystal structure, and thermal behavior of the first three-dimensional titanium(IV) diphosphonate with an open structure: Ti3O2(H2O)2(O3P-(CH2)-PO3)2·(H2O)2, or MIL-22. Inorganic Chemistry. 1999;**38**:5370-5373. DOI: 10.1021/IC990345M/SUPPL_FILE/IC990345M_S.PDF

[82] Nguyen HL. Perspectives on titanium-based metal-organic frameworks. Journal of Physics: Energy. IOP Publishing Ltd. April 1, 2021:021003. DOI: 10.1088/2515-7655/abe3c9

[83] Fang WH, Zhang L, Zhang J. Synthetic strategies, diverse structures and tuneable properties of polyoxo-titanium clusters. Chemical Society Reviews. 2018;**47**:404-421. DOI: 10.1039/C7CS00511C

[84] Li Y, Yang B, Liu B. MOF assisted synthesis of TiO2/Au/Fe2O3 hybrids with enhanced photocatalytic hydrogen production and simultaneous removal of toxic phenolic compounds. Journal of Molecular Liquids. 2021;**322**:114815. DOI: 10.1016/J.MOLLIQ.2020.114815

[85] Chen X, Peng X, Jiang L, Yuan X, Yu H, Wang H, et al. Recent advances in titanium metal–organic frameworks and their derived materials: Features, fabrication, and photocatalytic applications. Chemical Engineering Journal. 2020;**395**:125080. DOI: 10.1016/J.CEJ.2020.125080

[86] Qiu LG, Xu T, Li ZQ, Wang W, Wu Y, Jiang X, et al. Hierarchically micro- and mesoporous metal–organic frameworks with tunable porosity. Angewandte Chemie International Edition. 2008;**47**:9487-9491. DOI: 10.1002/ANIE.200803640

[87] Sun D, Li Z. Robust Ti- and Zr-based metal-organic frameworks for Photocatalysis. Chinese Journal of Chemistry. 2017;**35**:135-147. DOI: 10.1002/CJOC.201600647

[88] Brozek CK, Dincă M. Ti3+–, V2+/3+–, Cr2+/3+–, Mn 2+–, and Fe2+–substituted MOF-5 and redox reactivity in Cr- and Fe-MOF-5. Journal of the American Chemical Society. 2013;**135**:12886-12891. DOI: 10.1021/JA4064475/SUPPL_FILE/JA4064475_SI_001.PDF

[89] Nguyen HL, Vu TT, Le D, Doan TLH, Nguyen VQ, Phan NTS. A titanium-organic framework: Engineering of the band-gap energy for photocatalytic property enhancement. ACS Catalysis. 2017;**7**:338-342. DOI: 10.1021/ACSCATAL.6B02642/SUPPL_FILE/CS6B02642_SI_001.PDF

[90] Visible-Light-Active Photocatalysis: Nanostructured Catalyst Design, Mechanisms, and Applications | Wiley. n.d. https://www.wiley.com/en-ie/Visible+Light+Active+Photocatalysis:+Nanostructured+Catalyst+Design,+Mechanisms,+and+Applications-p-9783527342938 [Accessed: July 24, 2022]

[91] Hendon CH, Tiana D, Fontecave M, Sanchez C, D'Arras L, Sassoye C, et al. Engineering the optical response of the titanium-MIL-125 metal-organic framework through ligand functionalization. Journal of the American Chemical Society. 2013;**135**:10942-10945. DOI: 10.1021/ja405350u

[92] Canivet J, Fateeva A, Guo Y, Coasne B, Farrusseng D. Water adsorption in MOFs: Fundamentals and applications. Chemical Society Reviews. 2014;**43**:5594-5617. DOI: 10.1039/C4CS00078A

[93] Burtch NC, Jasuja H, Walton KS. Water stability and adsorption in metal-organic frameworks. Chemical Reviews. 2014;**114**:10575-10612. DOI: 10.1021/cr5002589

[94] Wang C, Liu X, Keser Demir N, Chen JP, Li K. Applications of water stable metal–organic frameworks. Chemical Society Reviews. 2016;**45**:5107-5134. DOI: 10.1039/C6CS00362A

[95] Wang K, Lv XL, Feng D, Li J, Chen S, Sun J, et al. Pyrazolate-based Porphyrinic metal-organic framework with extraordinary base-resistance. Journal of the American Chemical Society. 2016;**138**:914-919. DOI: 10.1021/JACS.5B10881/SUPPL_FILE/JA5B10881_SI_002.CIF

[96] Zhang G, Lan ZA, Wang X. Conjugated polymers: Catalysts for photocatalytic hydrogen evolution. Angewandte Chemie International Edition. 2016;**55**:15712-15727. DOI: 10.1002/ANIE.201607375

[97] Yu J, Low J, Xiao W, Zhou P, Jaroniec M. Enhanced photocatalytic CO2-reduction activity of anatase TiO2 by Coexposed {001} and {101} facets. Journal of the American Chemical Society. 2014;**136**:8839-8842. DOI: 10.1021/JA5044787/SUPPL_FILE/JA5044787_SI_001.PDF

[98] Liao X, Wei W, Zhou Y, Zhang M, Cai Y, Liu H, et al. A Ti-based bi-MOF for the tandem reaction of H2O2 generation and catalytic oxidative desulfurization. Catalysis Science & Technology. 2020;**10**:1015-1022. DOI: 10.1039/C9CY02318F

[99] Li SL, Xu Q. Metal–organic frameworks as platforms for clean energy. Energy & Environmental Science. 2013;**6**:1656-1683. DOI: 10.1039/C3EE40507A

[100] Han SY, Pan DL, Chen H, Bu XB, Gao YX, Gao H, et al. A Methylthio-functionalized-MOF Photocatalyst with high performance for visible-light-driven H2 evolution. Angewandte Chemie

International Edition. 2018;**57**:9864-9869. DOI: 10.1002/ANIE.201806077

[101] Isaka Y, Kondo Y, Kawase Y, Kuwahara Y, Mori K, Yamashita H. Photocatalytic production of hydrogen peroxide through selective two-electron reduction of dioxygen utilizing amine-functionalized MIL-125 deposited with nickel oxide nanoparticles. Chemical Communications. 2018;**54**:9270-9273. DOI: 10.1039/C8CC02679C

[102] Fu Y, Yang H, Du R, Tu G, Xu C, Zhang F, et al. Enhanced photocatalytic CO2 reduction over Co-doped NH2-MIL-125(Ti) under visible light. RSC Advances. 2017;**7**:42819-42825. DOI: 10.1039/C7RA06324E

[103] Serre C, Férey G. Hydrothermal synthesis and structure determination from powder data of new three-dimensional titanium(IV) diphosphonates Ti(O3P-(CH2)n-PO3) or MIL-25n (n=2, 3). Inorganic Chemistry. 2001;**40**:5350-5353. DOI: 10.1021/IC010260X/ASSET/IMAGES/MEDIUM/IC010260XN00001.GIF

[104] Ratnamala A, Reddy GD, Noorjahaan M, Manjunatha H, Janardan S, Kumar NS, et al. Titanium-based metal-organic frameworks for photocatalytic applications. Metal-Organic Frameworks for Chemical Reactions. 2021:37-63. DOI: 10.1016/B978-0-12-822099-3.00003-4

[105] Cassaignon S, Koelsch M, Jolivet JP. Selective synthesis of brookite, anatase and rutile nanoparticles: Thermolysis of TiCl4 in aqueous nitric acid. Journal of Materials Science. 2007;**4216**(42):6689-6695. DOI: 10.1007/S10853-007-1496-Y

[106] Zhu J, Li PZ, Guo W, Zhao Y, Zou R. Titanium-based metal–organic frameworks for photocatalytic applications. Coordination Chemistry Reviews. 2018;**359**:80-101. DOI: 10.1016/j.ccr.2017.12.013

[107] Yuan S, Liu TF, Feng D, Tian J, Wang K, Qin J, et al. A single crystalline porphyrinic titanium metal–organic framework. Chemical Science. 2015;**6**:3926-3930. DOI: 10.1039/C5SC00916B

[108] Hu S, Liu M, Li K, Zuo Y, Zhang A, Song C, et al. Solvothermal synthesis of NH2-MIL-125(Ti) from circular plate to octahedron. CrystEngComm. 2014;**16**:9645-9650. DOI: 10.1039/C4CE01545B

[109] Han Y, Han L, Zhang L, Dong S. Ultrasonic synthesis of highly dispersed Au nanoparticles supported on Ti-based metal–organic frameworks for electrocatalytic oxidation of hydrazine. Journal of Materials Chemistry A. 2015;**3**:14669-14674. DOI: 10.1039/C5TA03090K

[110] Nguyen HL, Gándara F, Furukawa H, Doan TLH, Cordova KE, Yaghi OM. A titanium-organic framework as an exemplar of combining the chemistry of metal- and covalent-organic frameworks. Journal of the American Chemical Society. 2016;**138**:4330-4333. DOI: 10.1021/JACS.6B01233/SUPPL_FILE/JA6B01233_SI_001.PDF

[111] Brozek CK, Dincă M. Cation exchange at the secondary building units of metal–organic frameworks. Chemical Society Reviews. 2014;**43**:5456-5467. DOI: 10.1039/C4CS00002A

[112] Kim M, Cahill JF, Fei H, Prather KA, Cohen SM. Postsynthetic ligand and cation exchange in robust metal-organic frameworks. Journal of the American Chemical Society. 2012;**134**:18082-18088.

DOI: 10.1021/JA3079219/SUPPL_FILE/JA3079219_SI_001.PDF

[113] McNamara ND, Hicks JC. Chelating agent-free, vapor-assisted crystallization method to synthesize hierarchical microporous/mesoporous MIL-125 (Ti). ACS Applied Materials & Interfaces. 2015;7:5338-5346. DOI: 10.1021/AM508799D/SUPPL_FILE/AM508799D_SI_001.PDF

[114] Shen L, Luo M, Huang L, Feng P, Wu L. A clean and general strategy to decorate a titanium metal-organic framework with noble-metal nanoparticles for versatile photocatalytic applications. Inorganic Chemistry. 2015;**54**:1191-1193. DOI: 10.1021/IC502609A/SUPPL_FILE/IC502609A_SI_001.PDF

[115] Fu Y, Sun L, Yang H, Xu L, Zhang F, Zhu W. Visible-light-induced aerobic photocatalytic oxidation of aromatic alcohols to aldehydes over Ni-doped NH2-MIL-125(Ti). Applied Catalysis B: Environmental. 2016;**187**:212-217. DOI: 10.1016/J.APCATB.2016.01.038

[116] Wang H, Yuan X, Wu Y, Chen X, Leng L, Zeng G. Photodeposition of metal sulfides on titanium metal-organic frameworks for excellent visible-light-driven photocatalytic Cr(vi) reduction. RSC Advances. 2015;**5**:32531-32535. DOI: 10.1039/c5ra01283j

[117] Nguyen HL. The chemistry of titanium-based metal–organic frameworks. New Journal of Chemistry. 2017;**41**:14030-14043. DOI: 10.1039/C7NJ03153J

[118] Batten SR, Robson R. Interpenetrating nets: Ordered, periodic entanglement. Angewandte Chemie International Edition. 1998;**37**:1460-1494. DOI: 10.1002/(SICI)1521-3773(19980619)37:11<1460::AID-ANIE1460>3.0.CO;2-Z

[119] Jiang F, Dai L, Shi Y, Wang Z. A three-dimensional chiral crystal structure constructed from a chiral triazolate ligand showing an SrSi2 topology: Poly[bis(μ 3-3,5-diethyl-1,2,4-triazolato-κ3N1: N2: N4)trisilver nitrate]. Acta Crystallographica Section C: Crystal Structure Communications. 2013;**69**:1485-1487. DOI: 10.1107/S0108270113029788/YF3050ISUP2.HKL

[120] Bueken B, Vermoortele F, Vanpoucke DEP, Reinsch H, Tsou C-C, Valvekens P, et al. A flexible photoactive titanium metal–organic framework based on a [TiIV3(μ3-O)(O)2(COO)6] cluster. Angewandte Chemie. 2015;**127**:14118-14123. DOI: 10.1002/ANGE.201505512

[121] Hong K, Bak W, Chun H. Unique coordination-based heterometallic approach for the stoichiometric inclusion of high-valent metal ions in a porous metal-organic framework. Inorganic Chemistry. 2013;**52**:5645-5647. DOI: 10.1021/IC400607W/SUPPL_FILE/IC400607W_SI_002.PDF